Clinicians' & Educator
on the Integrative Health & ... ...essions

Third Edition

Developed by
Academic Collaborative for Integrative Health

Partner Organizations

Alliance for Massage Therapy Education
Association of Accredited Naturopathic Medical Colleges
Association of Chiropractic Colleges
Association of Midwifery Educators
Council of Colleges of Acupuncture and Oriental Medicine

Project Managers and Editors

Renee Motheral Clugston
Elizabeth Goldblatt, PhD, MPA/HA
Beth Rosenthal, MPH, MBA, PhD
Pamela Snider, ND
John Weeks

www.integrativehealth.org

Acknowledgements

We gratefully acknowledge our partner organizations and the authors and editors of these chapters, the ACIH Board, Working Group, Council of Advisors, Organizational, College, and Associate members, ACIH staff, and our many national colleagues. The book's original design, layout, printing, marketing and distribution were made possible by a generous grant from the National University of Health Sciences.

We gratefully acknowledge the CEDR First and Second Edition Project Managers and Editors: Elizabeth Goldblatt, PhD, MPA/HA, Sheila Quinn, Beth Rosenthal, MPH, MBA, PhD, Pamela Snider, ND, & John Weeks. In addition, we wish to thank Deb Hill, MS, PERL Project Manager, for her thorough proofreading of this current version.

Clinicians' and Educators' Desk Reference on the Integrative Health and Medicine Professions

PO Box 1432
Mercer Island, WA 98040
www.integrativehealth.org

Disclaimer: This publication is intended for use as an educational tool only and should not be used to make clinical decisions concerning patient care. Information is subject to change as educational and regulatory environments evolve. In addition, the role of the partner organizations was to identify authors. These organizations have not endorsed the content of the book, and their participation in selecting the chapter authors was not intended as an endorsement.

ISBN 978-1-387-10157-3

"The *CEDR* advances the understanding of and appreciation for integrative health professions and disciplines, enhancing our capacity to provide the best available care for our patients. This is a useful text that will be employed frequently in the day-to-day care of patients, and will certainly remain a valued reference for each of our students long after graduation."- ***Joseph E. Brimhall, DC,*** *President, University of Western States; Chair, Northwest Commission on Colleges and Universities; Chair, Oregon Collaborative for Integrative Medicine*

"I urge my class of fourth-year medical students and residents to refer to the *CEDR* often in the future. It really is a wonderful, IPE/C-themed guide." - ***Shelley R. Adler, PhD****, Professor, Department of Family and Community Medicine; Director of Education, Osher Center for Integrative Medicine University of California, San Francisco*

"The *CEDR* certainly fills a hole in health education textbook resources...healthcare students of all disciplines need this concise and current summary of this evolving and exciting sector of the health care system. It is well-organized, up-to-date and will certainly be used to foster greater inter-professional understanding and inter-disciplinary behavior among healthcare practitioners." ***-William Meeker, DC, MPH****, President, Palmer College-West Campus*

"The data and perspectives included in the book will be especially useful to me and I suspect that this book will be useful to all involved in the diverse professions represented by the team that produced this book. Your team did a superb job in assembling this fine book." ***-Lawrence J. Ryan, PhD,*** *President, Yo San University of Traditional Chinese Medicine*

"Even though I consider myself quite well-informed about this topic, I was impressed with how much I learned from this desk reference...This book will help facilitate greater and more respectful collaboration among practitioners from multiple disciplines." ***-Bill Manahan, MD****, Assistant Professor Emeritus, Department of Family Medicine and Community Health; University of Minnesota Academic Health Center, Minneapolis, Minnesota*

"This reference can be helpful when responding to patient questions regarding other disciplines, as well as when referring patients to those same disciplines. We know what we do as naturopathic physicians, but what do we know about the details of those other approaches that we find patients also utilizing? Having additional information about those approaches might help us to better select which one is a better fit for referral for our patients." -***Marcia Prenguber, ND***, *Dean, College of Naturopathic Medicine, University of Bridgeport*

"This is a reference book every student interested in IHM should have on their shelf." - ***Adam Perlman, MD, MPH,*** *Program Director, Duke Leadership Program in Integrative Health; CEO & Co-Founder of Journey Wellness*

"What we are learning from our fellow health care providers is that we need the information provided by the *CEDR* to educate current and future providers and to develop the collaborative teams so desperately needed for the changing health care needs of our nation. " -***Stacy Gomes, EdD***, *Vice President of Academic Affairs, Pacific College of Oriental Medicine*

"Offers all of us, for the first time, a clear view of the IHM professions: their skills, approaches to patient care, education and accreditation… extremely useful for current health practitioners who need to know more about each of the IHM disciplines, and for consumers who are evaluating their healthcare choices." -***Elaine Zablocki***, *Townsend Letter*

"This well-referenced book will appeal to those clinicians with a scientific- and evidence-based mindset, and yet is written in simple enough language so that no therapist, or client, would be intimidated by it... I think it's important to get this book into the hands of as many mainstream-medicine practitioners as possible. "- ***Laura Allen***, *In Massage Today*

"Congratulations on producing a very useful and much needed reference." - ***Aviad Haramati, PhD***, *Professor, Integrative Physiology of the Department of Biochemistry, Molecular & Cellular Biology at Georgetown University School of Medicine; Founding Vice Chair, Academic Consortium for Integrative Medicine and Health*

"Safe and effective healthcare depends upon team members' mutual respect and understanding, and team-based care is certainly the way of the future. The *CEDR* is an invaluable resource in that it provides a consistent message to all healthcare workers about the increasingly important role of the complementary professions. Every healthcare provider should have a copy." -***Michael Wiles, DC, MEd***, *Dean, Keiser University College of Chiropractic*

"The *CEDR* is an important tool in the education of medical students, faculty and staff in all medical schools. It brings cohesion to the fields of all practitioners, and allows non-medical people to understand the paradigms of each kind of medicine, and how they can work together for the health of patients." ***-Rita Bettenburg, ND****, Past Dean, Naturopathic Medicine, National University of Natural Medicine*

"This handbook is an excellent resource for all nurses. It incorporates information about the licensed IHM Professions including their Philosophy, Characteristics, Scope of Practice, Education, Regulation and Certification, Research, Challenges and Opportunities, and Resources." ***-American Holistic Nurses Association****, News From AHNA, Vol. 9 No. 3*

"The desk reference is intended for clinicians interested in collaboration and referral and will also serve as a textbook and useful resource for educators and students." ***-Midwifery Education Accreditation Council*** *in MANA NEWS (Midwives Alliance of North America)*

"The *CEDR* will be a required book for each student that goes through our program. We are using the *CEDR* to help our students understand how to integrate their services with the services other complementary and integrative health professions offer." ***-Stan Dawson DC, LMBT****, Vice President, Alliance for Massage Therapy Education (AFMTE)*

"This book is a great platform for developing collaborative relationships to serve our patients." ***-Andrew Weil, MD****, Author; Founder and Director, Arizona Center for Integrative Medicine*

Table of Contents

Continued on next page

Appendices

Preface

Chiropractic, massage, acupuncture and Oriental medicine, midwifery, and naturopathic medicine were in common use before national surveys documented their sizeable footprint within the US healthcare delivery system.

Five national surveys conducted since 1990 have found that at least a third of US adults routinely use complementary and integrative health and medicine (CIHM, previously known as 'CAM') therapies to treat their principal medical conditions. All five surveys documented the fact that each year Americans schedule hundreds of millions of office visits to licensed CIHM professionals at a cost of tens of billions of dollars, most of which is paid for out-of-pocket. The most recent national survey published by the Centers for Disease Control (NHIS 2012) reported that out-of-pocket expenditures associated with CIHM practices account for approximately 9.2% of all out-of-pocket health care expenditures in the United States. The discovery of this "hidden mainstream" of American health care is no longer disputed.

These national surveys have also confirmed that the majority of patients who use CIHM therapies generally do not disclose or discuss them with their primary care physicians or subspecialists. This lack of interdisciplinary communication is not in a patient's best interest. Making matters more complicated, it is also known that among individuals who use CIHM therapies, the majority tend to use multiple CIHM practices as opposed to a single modality to treat their medical problems. Regrettably, many individuals seek combinations of conventional and complementary therapies in a non-coordinated rather than an integrated fashion.

The Institute of Medicine, in its report *Complementary and Alternative Medicine in the United States*, has helped set the stage for future research in this area. The report states:

> Studies show that patients frequently do not limit themselves to a single modality of care – they do not see complementary and alternative medicine (CAM) and conventional medicine as being

> mutually exclusive – and this pattern will probably continue and may even expand as evidence of therapies' effectiveness accumulate. Therefore, it is important to understand how CAM and conventional medical treatments (and providers) interact with each other and to study models of how the two kinds of treatments can be provided in coordinated ways. In that spirit, there is an urgent need for health systems research that focuses on identifying the elements of these integrative medical models, their outcomes and whether these are cost effective when compared to conventional practice.
>
> —Institute of Medicine
> *Report on Complementary and Alternative Medicine in the US*
> National Academy of Sciences, 2005

In this era of cost containment and healthcare reform, in addition to the development of patient centered medical homes and accountable care organizations, this recommendation needs to become a priority. In order to design and implement such studies, however, we must first define and describe each of the relevant CIHM modalities and professional groups in order to establish standards that lend themselves to reproducible research. Moreover, referrals between professional communities and the training of multi-disciplinary teams consisting of both conventional and complementary care practitioners cannot begin unless and until all participating professional groups learn more about one another.

This *Clinicians' and Educators' Desk Reference on the Integrative Health and Medicine Professions* contributes enormously to the goal of designing, testing, and refining models of multidisciplinary, comprehensive, integrative care. It thoughtfully yet concisely describes each of the major CIHM professions. This information will be useful to patients, healthcare professionals, educators, students, and those responsible for future clinical research and healthcare policy. The Academic Collaborative for Integrative Health is to be commended for making this information readily available.

There is a Chinese proverb which reads, "The methods used by one will be faulty. The methods used by many will be better." Over the next decade, let us test this idea, collaboratively and with all the scientific rigor, clinical expertise, and integrity we have to offer. The next generation will surely thank us.

David Eisenberg, MD
Adjunct Associate Professor, Harvard T. H. Chan School of Public Health
Former Chair, Division of Research and Education in Complementary and Integrative Medical Therapies, Harvard Medical School

Foreword

This handbook is arriving on the healthcare scene in the midst of tumultuous and highly politicized debate about sustainable reforms in the US health system, escalating insurance costs, and increasingly limited choices of plans. As such, it contributes to the dialogue in a substantive way by providing, logically and clearly, well organized descriptions of the major complementary and integrative healthcare professions.

Three issues received considerable attention in an Institute of Medicine (now named Health and Medicine Division) Summit on Integrative Medicine and the Health of the Public (February 2009) that remain viable and crucial as we attempt to move our health systems forward:

- The US has an inadequate primary care workforce. Increasingly, there is recognition that there are licensed Complementary and Integrative Health and Medicine (CIHM, previously known as 'CAM') practitioners who are prepared to be the first point of entry for many patients. There are also advanced practice nurses who can serve as primary care providers. Current workforce planning needs to take into account a broader range of providers.
- There is a clear need to reorient the system from one that focuses on disease to one that also promotes health. CIHM providers have much to contribute in both health promotion and disease treatment.
- Given the complexity of disease and the multiple challenges to human health, rarely can one provider meet all of the physical, emotional, psychological, and spiritual needs of a patient. A team approach that takes into account the capacities of a full range of providers, including CIHM practitioners, is critical.

We believe that the time is right for a well-planned and deliberate strategy to integrate CIHM practitioners into the nation's healthcare system. This will require legislative commitment, action, and financial models showing a strong cost-benefit outcome.

CIHM providers are well positioned to focus on prevention, lifestyle change, health coaching, and management of certain chronic diseases and painful conditions. The latter problem, chronic pain, has become especially critical in our era as evidenced by an opiate addiction epidemic and countless lives affected. Many of these patients do not respond to or are harmed by conventional medications and procedures aimed at controlling their chronic pain. Systematic integration of CIHM providers has the potential to increase quality of care of chronic pain and other conditions while decreasing costs of health care.

This handbook offers an orientation for the healthcare teams of the future. Many conventional providers simply do not know enough about the training, scope of practice, and evidence supporting many of the complementary and integrative therapies and practitioners described here. This book helps to bridge that information gap.

In the future, we would like to see more practical and detailed descriptions of interdisciplinary team collaborations between physicians, advanced practice nurses, physician assistants, and physical therapists from the conventional medical world and the CIHM practitioners listed in this handbook. The processes of facilitating referral, interdisciplinary communication, and team building must be built into the training of medical and nursing schools for the next phase of evolution into true interdisciplinary and transdisciplinary healthcare teams.

As members of the Academic Consortium for Integrative Medicine and Health, our work has involved bringing mainstream medicine, nursing, and allied health professionals into a positive relationship with complementary and integrative therapies, disciplines, and professionals. Progress is occurring gradually though much remains to be done in these areas including policy changes regarding reimbursement and curriculum design.

However, the benefit of a handbook such as this one is that it creates a common ground in which patient-centered care is central and in which respect for others with professional skills and perspectives different from our own is encouraged. We see this type of publication as a helpful textbook and reference not only to the CIHM community but to the conventional healthcare professions as well.

Victor S. Sierpina, MD, ABFM, ABIHM
Past Chair, Academic Consortium for Integrative Medicine and Health
Director, Medical Student Education, WD and Laura Nell Nicholson Family
Professor of Integrative Medicine
Professor, Family Medicine
University of Texas Distinguished Teaching Professor
University of Texas Medical Branch

Mary Jo Kreitzer, PhD, RN, FAAN
Past Vice-Chair, Academic Consortium for Integrative Medicine and Health
Director, Center for Spirituality and Healing
Professor of Nursing
University of Minnesota

Adam Perlman, MD, MPH
Past Chair, Academic Consortium for Integrative Medicine and Health
Program Director, Duke Leadership Program in Integrative Health
CEO & Co-Founder of Journey Wellness

April 2017

Introduction: Deepening Engagement in Interprofessional Practice and Education

Third Edition Authors/Editors: Elizabeth Goldblatt, PhD, MPA/HA; David O'Bryon, JD, CAE; Beth Rosenthal, PhD, MBA, MPH

First and Second Edition Authors: John Weeks and Elizabeth Goldblatt, PhD, MPA/HA

This book is intended for use by educators, healthcare professionals, students and administrators, and consumers of health care as well as all those involved in and passionate about changing the current healthcare system to one that is collaborative, team-based, patient-centered and focused on disease prevention, health and well-being.

ACIH began in 2004 as the Academic Consortium for Complementary and Alternative Health Care (ACCAHC). Our new name – the Academic Collaborative for Integrative Health, also referred to as 'ACIH' or 'The Collaborative' – reflects our growth, recognition and involvement with and in the larger healthcare system. In addition to a new name we have a new website address - www.integrativehealth.org.

The use of the phrases integrative health and medicine (IHM) and integrative health (IH) describe the disciplines that ACIH represents. These two acronyms are used interchangeably throughout this book in place of 'CAM' (complementary & alternative medicine), which no longer reflects the role of the IHM health professionals in our society.

We publish this third edition as an update of the second edition, which came out in 2013, to keep the information current and in the context of the movement towards interprofessional practice and education.

Vision

ACIH envisions a healthcare system that is multidisciplinary and enhances competence, mutual respect and collaboration across all healthcare disciplines. This system will deliver effective care that is patient centered, focused on health creation and healing, and readily accessible to all populations.

Mission

ACIH enhances health by cultivating partnerships and advancing interprofessional education and collaborative practice (IPE/CP).

The following statement from the Lancet report entitled "Health Professionals for a New Century" published in 2010, reflects ACIH's mission and vision as well as our emphasis on the importance of IPE/CP.[1]

> "Our goal is to encourage all health professionals, irrespective of nationality and specialty, to share a common global vision for the future. In this vision, all health professionals in all countries are educated to mobilize knowledge, and to engage in critical reasoning and ethical conduct, so that they are competent to participate in patient-centered and population-centered health systems as members of locally responsive and globally connected teams. The ultimate purpose is to assure universal coverage of high quality comprehensive services that are essential to advancing the opportunity for health equity within and between countries. The aspiration of good health commonly shared, we believe, resonates with young professionals who see value and meaning in their work."

ACIH is pleased to be involved with multiple national and international projects related to this promotion of inclusive, collaborative, team-based care that is patient-centered. These include: development of appropriate competencies for education, research and practice in integrative settings; sponsorship of, and participation in, the National Academies of Sciences, Engineering, and Medicine's Health and Medicine Division (previously known as the Institute of Medicine) Global Forum on Innovation in Health Professional Education; having representation on the leadership team of the National Center for Integrative Primary Healthcare housed at the

University of Arizona; involvement in the Collaboration Across Borders conferences; active engagement in a national campaign to promote an integrated approach to pain treatment; and most recently, participation in the Center for Interprofessional Practice and Education annual summit at the University of Minnesota.

Our organization represents the five licensed integrative health and medicine (IHM) professions as well as some of the traditional world medicines and emerging professions. Providers from these various disciplines work in community clinics, conventional medicine out-patient clinics and hospitals, private and integrative clinics, may have insurance and healthcare benefit designs, and are engaged in research and education grants, government programs, and academic centers throughout the United States. The five licensed IHM professions are: acupuncture and Oriental medicine, chiropractic, direct-entry midwifery, massage therapy, and naturopathic medicine. These professions have all achieved an advanced level of regulatory maturity, including attaining the important professional benchmark of earning United States Department (Secretary) of Education recognition for the agencies that accredit their educational institutions. These five IHM disciplines represent roughly 436,000 practitioners nationwide.

In order to support expanded understanding, informed referrals, and choices across the entire integrative healthcare practice domain, we also offer information on the related fields of advanced nutrition, Ayurvedic medicine, holistic nursing, homeopathy, integrative health & medicine, integrative medicine, and yoga therapy. Together, practitioners from the licensed IHM disciplines and related fields represent the integrative healthcare professionals whom a practitioner is most likely to encounter in developing a more integrated team of community resources for patients.

We are pleased to witness the excitement and acclaim with which this book is being received by many educators, students and providers across multiple disciplines. The CEDR is freely available in a downloadable PDF format, and hard copies are available for purchase at www.integrativehealth.org/desk-reference-integrative-health-prof essions/.

This book offers comprehensive descriptions of each of the ACIH and related disciplines and the chapters are authored by individuals with expertise in their subject matter, chosen by their disciplines' national organizations. We believe that it is vitally important that all health professionals are well informed about each other and are trained to work collaboratively for the benefit of the patient.

Supporting Multidisciplinary Care: A Focus on Team-based Care

In this era in which the need to focus on wellness, well-being, and health promotion as well as disease prevention and better treatment of chronic diseases dominates the healthcare reform dialogue, multi-disciplinary teams form the foundation of most of the significant new and encouraging clinical models. A team approach to health care provides higher patient satisfaction and better health care, and reduces costs. These three areas are often referred to as the 'Triple Aim'.[2] Advocates of the Chronic Care Model, of prospective health care, of the medical home, of integrative medicine, and of functional medicine, all promote an increasingly multidisciplinary medicine and healthcare approach. The movements for interprofessional practice and education and for integrated pain care are similarly focused. Success in such teams begins with quality information about the knowledge, skills and background of potential contributors from diverse health professions. It is interesting to note that virtually all conventional medicine and integrative healthcare accreditation commissions now require some level of formal training in interprofessional education and clinical practice (IPE/CP). The Interprofessional Educational Collaborative (IPEC) recently updated its competencies, reflecting the growth and need for collaboration.[3]

Health professionals can still tend to be educated in silos. We have a saying in the Collaborative that "those who are educated together practice together." If we wish to develop a patient-centered healthcare system, then we must break out of these isolated silos and learn more about one another as well as learn to work in collaborative, team-based, patient-centered care.

A Collaborative Project

In this book, ACIH worked with key national educational organizations to select authors whose expertise and experience as clinicians and educators would inform the writing. We believe that in this day and age when patients are putting together their own healthcare teams, that all healthcare professionals have the ethical responsibility to become more educated about the IHM fields.

Past publications relative to our disciplines focused mainly on individual agents, modalities, or therapies. Insufficient attention was given to exploration of the whole practices of these disciplines and the systems of care that the distinctly licensed IHM practitioners are educated to provide. Too often, the portraits of these professions, when they are offered, came from outside observers rather than from those within in the IHM disciplines.

The historically limited focus of many texts discussing the IHM disciplines often did a disservice to educators, practitioners and consumers. A description of a single natural therapy, for instance, does not teach about the context in which it is offered by an actual practitioner. IHM disciplines reflect whole systems of care rather than more limited modalities or therapies. A modality or therapeutic orientation (in contrast to a focus on the discipline as a whole) falls short in assisting students, educators, and clinicians to understand how to communicate with and work in teams with other practitioners.

In a study of integrative health and medicine definitions, Rosenthal and Lisi concluded: "using the terms modality/therapy/intervention/treatment rather than the terms professional or discipline discounts the importance of delivering the modality/therapy/intervention/treatment in context of its discipline, by practitioners who are educated and trained in the discipline. Leaving disciplines and health professionals out of the definition effectively leaves out the rich experience and context of the discipline..."[4]

This text provides our disciplines' overall philosophy, distinct approach to patients, depth of the educational training, accreditation requirements, regulatory status, scope of practice, research, and references. The healthcare disciplines portrayed in this book tend also to emphasize preventive medicine, health and well-being, the practitioner-patient relationship, the promotion of wellness, and the teaching of self-care practices in addition to the treatment of various diseases and conditions. As became abundantly clear through the ACIH project "Meeting the Nation's Primary Care Needs", IHM practitioners tend also to be "first choice" or "first access" providers for the patients who come to them.[5]

Additional Context for this Book

We have taken a perspective that is part of a broader shift in the integrative healthcare practice that is now focusing on interprofessional education and collaborative practice. In 2005, the Academic Consortium for Integrative Medicine and Health (ACIHM, formerly known as the Consortium of Academic Health Centers for Integrative Medicine), changed its operating definition of integrative medicine. The amended definition emphasizes the importance of integrating not only "modalities" but also "healthcare professionals and disciplines." (See the Definition of Integrative Medicine at https://www.imconsortium.org/about/about-us.cfm) Notably, the interprofessional education movement stresses that "egalitarianism" is a core characteristic of the model that infuses optimal team care. All disciplines may not contribute equally to a patient's care but the relationship between them is one of respect for what each brings to the patient. This is a core ACIH value.

The commitment of educators from the five licensed disciplines featured in this book, along with our traditional world medicine and emerging professions colleagues, to come together and form ACIH as an independent organization is also evidence of the emerging multidisciplinary universe. While the artifact that had been called "CAM" seemed to neatly bundle these distinct disciplines into a group, in truth, these professions also emerged in separate silos and each offers a distinct approach to health care. We are entering an era that values knowledge and skills in multidisciplinary teams. Membership in ACIH reflects a commitment to developing new knowledge, skills, and attitudes that promote team-based collaboration and patient-centered care.

The research environment also offers evidence of this shift toward broader dialogue from the limited focus on the use of individual agents. Attendees of the 2009 North American Research Conference on Complementary and Integrative Medicine, the most significant research meeting for integrative practice, witnessed a noticeable increase in program content relative to whole systems of care, patient-centered outcomes in care that focus on the whole person, and the challenges in capturing outcomes of actual practices. The research conferences since 2009 also are focusing more on whole systems of care. In addition, this conference was expanded in 2016 to include not only research but also education, clinical care, and policy. These bi-annual conferences are now entitled the "International Congress on Integrative Medicine and Health."

The 2011-2015 Strategic Plan of the National Institutes of Health National Center for Complementary and Alternative Medicine (NCCAM; now called the National Center for Complementary and Integrative Health/NCCIH), marked a shift toward effectiveness research and real world outcomes. This relatively new approach is continued in the 2016-2020 Strategic Plan. The ACIH-NCCIH dialogue successfully supported elevation of the importance of researching the impact of healthcare disciplines on populations. (See ACIH Research Working Group Led Communication with NIH NCIH on the 2011-2015 Strategic Plan as well as on the current NIH NCCIH Strategic Plan for 2016-2020.) [6,7]

This refocusing of the integration dialogue is also reflected in the Patient Protection and Affordable Care Act (2010). That law directly includes "licensed complementary and alternative medicine practitioners", and "integrative health practitioners" and "integrative health" in sections related to workforce, medical homes, prevention and health promotion, patient-centered outcomes research and, indirectly, in a requirement for non-discrimination among provider types in health plans. The language establishing a new focus on health, wellness and prevention is aligned with the health orientation and wellness principles of the licensed integrative healthcare professions and related integrative practice disciplines.

We have organized this book to facilitate ready access to useful information. We continue in this introduction with some general information on the five licensed IHM professions. This leads directly into the chapters from each of these five licensed IHM professions. Similar information on the related integrative practice fields is found in later chapters. Additional information about ACIH is shared in the appendices. We include some accomplishments since ACIH was founded as a project of the Integrative Healthcare Policy Consortium (www.ihpc.org) in 2004. Given ACIH's organizational structure and focus on interprofessional education and practice, we provide a look at some of the collaborative, interprofessional activities with which we are proud to be involved.

Data on the Licensed Integrative Healthcare Professions

As noted above, the five licensed IHM professions have each distinguished themselves by achieving certain benchmarks of professional development. These benchmarks include creating councils of colleges or schools, establishing an accrediting agency, and then earning recognition by the US Department (Secretary) of Education. Each achievement marks a stage of maturation.

In addition, each of these licensed professions has also created a national certifying or testing agency, which is important to the public and public safety as well as to state regulators as the profession begins the long process of promoting licensing in each state. To expand licensing, initial groups of practitioners in each state usually formed state associations and created coalitions of supportive patients, consumers and professional colleagues to mobilize members of one legislature after another to enact licensing statutes for these professions.

Success in achieving these benchmarks is a prerequisite for integration into the broader health care arena in the US. Members of these disciplines are able to meet core credentialing standards for participation in conventional payment and delivery systems. Table I.1

provides data on the status of each profession in these benchmarking processes. As is clear, while all five disciplines have a federally recognized national specialty accrediting agency, the number of states in which licensing has been established varies significantly.

Table I.1

Development of Standards by the Licensed Integrative Healthcare Professions

Profession	Accrediting agency established	US Department of Education recognition	Recognized schools or programs	Standardized national exam created	State regulation *	Estimated number of licensed practitioners in the US
Acupuncture and Oriental medicine	1982	1988	57	1985	46 states + DC	28,000
Chiropractic Health Care^	1971	1974	15	1963	50 states + DC, Puerto Rico, and all other US territories /insular areas	77,000
Massage therapy	1982	2002	65**	1994	46 states, DC, Puerto Rico, Virgin Islands	325,000
Direct-entry Midwifery	1991	2001	10	1994	28 states	2,900
Naturopathic medicine	1978	2003	8	1986	19 states, DC, Puerto Rico, Virgin Islands	5,000

*For chiropractors, this category uniformly represents licensing statutes; for naturopathic physicians, this category primarily represents licensing statutes, with a few states offering registration statutes; for acupuncture, virtually all states use licensure; for massage and direct-entry midwifery, there is a mixture of licensing, certification, and registration statutes.

^Although the Council on Chiropractic Education was first incorporated in 1971, there were other accrediting agencies and activities within the chiropractic profession dating back to 1935.

**COMTA Accredited schools/programs. Does not include institutions accredited by non-specialized agency.

Partnerships with National Organizations for Content

Many questions arise when considering a publication to describe a number of unique professions. Who should be charged with authoring the chapters? How can we ensure balance and accuracy?

We chose to address these questions by working with and through authors, and in later editions content editors, selected by leading national educational organizations affiliated with the five licensed integrative and health professions. For the core chapters we contacted the organizational members of ACIH, and partnered with the profession's council of colleges and schools.

For the related integrative practice fields, we secured authors/ editors who were selected by the leading professional organizations in their respective fields. Table I.2 lists the chapter partner organizations, authors and editors. We are deeply grateful for all their contributions. Authors and editors are listed in alphabetical order in Table I.2, and in each chapter's third edition listings.

Table I.2

Partner Organizations and Authors

Chapters	Partner Organizations	Authors & Editors
Acupuncture and Oriental Medicine	Council of Colleges of Acupuncture and Oriental Medicine	Steve Given, DAOM, LAc Elizabeth Goldblatt, PhD, MPA/HA Mark McKenzie, PhD, LAc Vitaly Napadow, PhD, LAc Catherine Niemiec, JD, LAc David Sale, JD, LLM Kory Ward-Cook, PhD, MT (ASCP), CAE Jason Wright, MS, LAc, Dipl.OM
Chiropractic Health Care	Association of Chiropractic Colleges	Joseph Brimhall, DC David O'Bryon, JD, CAE Reed Phillips, DC, PhD Michael Wiles, DC, MEd, MS
Direct-entry Midwifery	Association of Midwifery Educators (2017 Edition) Midwifery Education Accreditation Council (2009 & 2013 editions)	Courtney L. Everson, PhD Jo Anne Myers-Ciecko, MPH Nichole Reding, MA, CPM, LDM
Massage Therapy	Alliance for Massage Therapy Education (2013 & 2017 editions) American Massage Therapy Association Council of Schools (2009 edition)	Stan Dawson, DC, LMBT Cherie Monterastelli, RN, MS, LMT Angie Myer, MA, HHP Jan Schwartz, MA, BCTMB Cherie Sohnen-Moe, BA Pete Whitridge, BA, LMT Kate Zulaski, BA, BCTMB
Naturopathic Medicine	Association of Accredited Naturopathic Medical Colleges	Christa Louise, PhD, MS Paul Mittman, ND, EdD Elizabeth Pimentel, ND, LMT Marcia Prenguber, ND Daniel Seitz, JD, EdD Michael Traub, ND, DHANP, FABNO Patricia Wolfe, ND JoAnn Yanez, ND, MPH, CAE
Related integrative practice fields		
Advanced Nutrition	Board for Certification of Nutrition Specialists	Jeffrey Blumberg, PhD, FASN, FACN, CNS-S Dana Reed, MS, CNS, CDN

Ayurvedic Medicine	National Ayurvedic Medical Association	Felicia Tomasko, RN
Holistic Nursing	American Holistic Nurses Association	Carole Ann Drick, PhD, RN, AHN-BC Carla Mariano, EdD, RN, AHN-BC, FAAIM
Homeopathy	Accreditation Commission for Homeopathic Education in North America (2013 & 2017 editions) American Medical College of Homeopathy (2009 edition)	Rick Cotroneo, MA, CCH Alastair Gray, BAHons MSC, DSH, PCH, PCHom Todd Rowe, MD, MD(H), CCH, DHt Heidi Schor
Integrative Health & Medicine	Academy of Integrative Health & Medicine (2017 edition) American Holistic Medical Association (2009 & 2013 editions)	Hal Blatman, MD Steve Cadwell Kjersten Gmeiner, MD Mimi Guarneri, MD Bradly Jacobs, MD Donna Nowak, CH, CRT Tabatha Parker, ND Molly Roberts, MD, MS
Integrative Medicine	Academic Consortium for Integrative Medicine and Health	Margaret A. Chesney, PhD Benjamin Kligler, MD, MPH Victor Sierpina, MD
Yoga Therapy	International Association of Yoga Therapists	John Kepner, MA, MBA, C-IAYT

Templates for the Chapters

The authors for the core chapters agreed to write approximately 6,000 words according to a template that was intended to make this a user-friendly reference guide for readers. For the related integrative practice fields, we used a similar template, but asked for 1,000 words. We gave the authors flexibility in the extent to which each would choose to focus on an individual section from among the following:

Philosophy, Mission, Goals
Characteristics and Data
Clinical Care
- Approach to patient care
- Scope of practice
- Referral practices
- Third-party payers

Integration activities
Education
- Schools and programs
- Curriculum content
- Faculty and other training information

Accreditation

Regulation and Certification

Regulatory status

Examinations and certifications

Research

Challenges and Opportunities

Resources

Organizations and websites

Bibliography

Citations

The authors understood that their chapters would have to be generally accepted within and outside borders of their own profession. The content was reviewed by an ACIH editing team.

Collaboration

We offer this book as a tool to facilitate greater understanding and collaboration between and among well-informed health professionals, educators and students, and to support better health care for patients. Respectful collaboration has been our vision and mission since ACIH was first conceived. The Appendices of this book shares some advances in cross-disciplinary education and practice over the past decade in which ACIH has been proud to be involved.

This book also represents collaboration with donors and with organizations that enhance our ability to fulfill our mission.

We are deeply grateful to Lombard, Illinois-based National University of Health Sciences (NUHS-www.nuhs.edu) for a significant project grant which allowed us to bring this multi-year project – the CEDR – to completion in the 2009 first edition of this text. ACIH also wishes to thank the Council of Colleges of Acupuncture and Oriental Medicine (www.ccaom.org) for a 2007 contribution that supported initial work on this project.

Critical financial support which nurtured ACIH through its gestational process has come from Lucy Gonda. Her vision and philanthropic support gave us our first major financial donation in

2003. Joining Gonda in the ACIH Sustaining Donors Group which supported relationship development and staff time for this book are the Westreich Foundation, the Leo S. Guthman Fund, NCMIC Foundation, Bastyr University and Life University. We thank, in particular, Ruth Westreich, Lynne Rosenthal, Lou Sportelli, DC, Daniel Church, PhD, and Guy Riekeman, DC respectively, for their decisions to move these organizations to support this work.

Finally, this book is now in its third edition. The publication of the third edition of this book comes amidst a new era of interprofessional respect and practice. More individuals, educators and health systems are realizing that optimal team care means widening the circle of involved health professionals. Thanks to each of you – the readers – for your own commitment to expanding your educational or clinical horizons to better understand the depth and breadth of these disciplines.

Citations

1. Frank J, Chen L, Bhutta Z, et al. Health professionals for a new century: transforming education to strengthen health systems in an interdependent world. *The Lancet*. 2010. Vol. 376(9756)1923-1958. doi: http://dx.doi.org/10.1016/S0140-6736(10)61854-5
2. The IHI Triple Aim. Institute for Healthcare Improvement website. http://www.ihi.org/engage/initiatives/tripleaim/pages/default.aspx
3. Interprofessional Education Collaborative. Core Competencies for Interprofessional Collaborative Practice: 2016 Update. Interprofessional Education Collaborative website. 2016. https://ipecollaborative.org/uploads/IPEC-2016-Updated-Core-Competencies-Report__final_release_.PDF.
4. Rosenthal B, Lisi A. A Qualitative analysis of various definitions of integrative medicine and health. *TIHC*. 2014, Vol. 5(4)ID: 5.4004
5. Goldstein M, Weeks J. *Meeting the Nation's Primary Care Needs: Current and Prospective Roles of Doctors of Chiropractic and Naturopathic Medicine, Practitioners of Acupuncture and Oriental Medicine, and Direct-Entry Midwives*. ACIH website. 2013. https://integrativehealth.org/primarycareproject

6. Menard M, Weeks J, Anderson B, et al. Consensus recommendations to NCCIH from research faculty in a transdisciplinary academic consortium for complementary and integrative health and medicine. *JACM.* 2015. Vol. 21(7)386–394. doi:10.1089/acm.2014.0295
7. NCCIH 2016 Strategic Plan. National Center for Complementary and Integrative Health website. 2016. https://nccih.nih.gov/about/strategic-plans/2016

Section I
Licensed Integrative Health and Medicine Professions

Acupuncture and Oriental Medicine

Chiropractic Health Care

Direct-Entry Midwifery

Massage Therapy

Naturopathic Medicine

Acupuncture and Oriental Medicine

Third Edition (2017) Editors:
Mark McKenzie, PhD, LAc; David Sale, JD, LLM;
Kory Ward-Cook, PhD, MT (ASCP), CAE;
Jason Wright, MS, LAc, Dipl.OM

Second Edition (2013) Editors:
Kory Ward-Cook, PhD, MT (ASCP), CAE;
Elizabeth Goldblatt, PhD, MPA/HA; David Sale, JD, LLM;
Vitaly Napadow, PhD, LAc; Catherine Niemiec, JD, LAc;
Steve Given, DAOM, LAc

First Edition (2009) Authors:
David Sale, JD, LLM; Steve Given, DAOM, LAc;
Catherine Niemiec, JD, LAc; Elizabeth Goldblatt, PhD, MPA/HA

Partner Organization: Council of Colleges of Acupuncture and Oriental Medicine (CCAOM)

About the Authors/Editors: Sale is Executive Director of the Council of Colleges of Acupuncture and Oriental Medicine. Given is the Dean of Clinical Education at the American College of Traditional Chinese Medicine at the California Institute of Integral Studies and a past member of the Accreditation Commission for Acupuncture and Oriental Medicine. Niemiec is President of the Phoenix Institute of Herbal Medicine and Acupuncture, and past Chair of the Accreditation Commission for Acupuncture and Oriental Medicine. Goldblatt is Past President of Council of Colleges of Acupuncture and Oriental Medicine, Doctorate of Acupuncture and Oriental Medicine Council and Faculty member at the American College of Traditional Chinese Medicine, and Executive Director of Academic Collaborative of Integrative Health. Ward-Cook is the Chief Executive Officer for the National Certification Commission for Acupuncture and Oriental Medicine. Napadow is Co-President of the Society for Acupuncture Research and Associate Professor at the Martinos Center for Biomedical Imaging at Massachusetts General Hospital and Harvard Medical School in Boston, MA, where he is also Director of the Center for Integrative Pain NeuroImaging. McKenzie is Executive Director of the Accreditation Commission for Acupuncture and Oriental Medicine and ACIH Board member. Wright is Director of Accreditation Services at the Accreditation Commission for Acupuncture and Oriental Medicine.

Philosophy, Mission, Goals

The history of acupuncture and Oriental medicine (AOM) extends back over 3000 years, with documentation of accumulated knowledge and experience appearing before the Han Dynasty (206 BCE to 220 CE), thus providing a long record of traditional use. Since the 1970s, AOM has experienced increasing popularity in the US.

The philosophy of this ancient therapeutic system is based on the concept of *qi* 气 (simplified) or 氣 (traditional), pronounced "chee", meaning energy/life force, and its flow through the body along channels or meridians. This ancient medicine, which can interface well with conventional medicine, has its own nomenclature, physiology, pathology, and therapeutics, which create a complex system of medicine documented in classical and modern texts.

Practitioners, educators, and national AOM organizations seek to promote a variety of complementary goals. These include: standards of excellence, competence, and integrity in the practice of the profession; public safety; excellence in AOM education; high quality health care; integration of AOM into the US healthcare system; chronic disease amelioration; a preventive health measure, especially as it impacts public health involvement; and assistance to communities affected by disaster, war, conflict, and poverty. Another goal of the AOM community of leaders is to provide evidence as to the efficacy and safety of acupuncture, Chinese herbology and Asian bodywork therapy through research and education.

Characteristics and Data

Prior to the 1970s, AOM was practiced in the US within Asian communities that arose after the completion of the Transcontinental Railroad in the 1840s. Acupuncture experienced a renaissance in the US in the early 1970s. Interest was heightened when James Reston, a newspaper columnist for the *New York Times,* wrote an article about the benefits of acupuncture he received during his recovery from an appendectomy in China when former President Nixon made his historic visit to China in 1971.[1] The following four major acupuncture institutions were formed in the early 1980s:

- Accreditation Commission for Acupuncture and Oriental Medicine (ACAOM), the US Department of Education-recognized national accrediting organization for AOM schools
- American Association of Acupuncture and Oriental Medicine (AAAOM), the national professional association of AOM Practitioners
- Council of Colleges of Acupuncture and Oriental Medicine (CCAOM), the voluntary national membership association for AOM schools
- National Certification Commission for Acupuncture and Oriental Medicine (NCCAOM), the national testing and certification organization for certifying competency of AOM practitioners.

There are approximately 27,000 to 29,000 licensed AOM practitioners throughout the US. The average (gross) income earned by licensed acupuncturists is $60,000; however, it is not uncommon for practitioners to earn in excess of this amount, with reported salaries in some instances exceeding $100,000.[2,3]

From 2009 – 2016, there were an average of 8,029 students enrolled in AOM graduate programs in the US.[4] In the early years of the profession in the US, students were primarily those looking for a second career. Increasingly, students today are looking to the AOM field as a first career. Although many of the graduates entering the AOM profession go into solo practice, there are increasing numbers in integrated healthcare delivery organizations, such as outpatient clinics, hospitals, and community clinics.

There is now expanded awareness of the recognition of acupuncture as a separate profession under the new federal Standard Occupational Classification (SOC) listing through the Bureau of Labor Statistics (BLS). The new code for Acupuncturists, SOC – 29-1291, will be included in the next edition of the BLS Occupational Handbook, which will be published in 2018.[5]

Clinical Care

Approach to patient care

Treatments provided by AOM practitioners identify a pattern or multiple patterns of disharmony within a patient and redress that disharmony in a variety of ways that may include any or all of the tools of Oriental medicine. AOM seeks to achieve balance by restoration of the harmonious movement of *qi* through the application of opposite energetic forces (e.g., clear heat through administration of cooling herbal formulas or acupuncture points specifically noted for their ability to release heat from the body) or through the strengthening of weak or deficient areas. The AOM practitioner typically proceeds by observing signs and symptoms that comprise a pattern of disharmony and one's constitutional state, rather than by developing a biomedical diagnosis based on etiologic assessments. AOM is used for chronic disease, prevention and wellness, and acute care. Treatments provided by AOM practitioners are appropriate for in-depth individual care as well as for simultaneous treatment of large groups of people, such as treatment of emergency care workers at the site of the 9/11 disaster and Hurricane Katrina, drug users in addiction recovery groups, and veterans suffering from PTSD. In addition, acupuncture is now utilized on the battlefield for pain control and a recent movement has seen many practitioners focus on delivering services in community treatment rooms.

Practitioners of AOM approach patients from a holistic perspective. They take into account all aspects of their health, recognizing the interconnectedness of the mind, emotions, and body, as well as the environment. When *qi* is stagnant or out of balance, which can result from natural, physical or emotional causes, illness and premature aging can occur. Oriental medicine involves a variety of techniques to restore balance to *qi* that has become stagnant or imbalanced, such as acupuncture and related therapies including:

- various needle techniques and tools
- cupping
- *gua sha* (or scraping technique)
- moxibustion
- modern use of electrical stimulation and cold laser therapy
- Asian forms of massage such as acupressure, shiatsu, and *tui na*
- herbal medicine
- exercise (*tai qi* and *qi gong*)
- Oriental dietary therapy
- meditation

Unique methods of diagnosis include the ability to interpret topography of the tongue, palpation of pulse and acupuncture points, and extensive observation and interpretation of all bodily symptoms and physical features according to Oriental medical physiology and diagnostic criteria. This medicine is used to treat a broad range of illnesses ranging from rehabilitation and pain management to virally mediated disorders and neurological complaints. There are several styles of Oriental medicine, including Traditional Chinese Medicine (TCM), Worsley/5-element theory, Japanese meridian theory, French Energetics, and Korean hand therapy.

Practitioners generally spend a significant amount of time developing a collaborative relationship with their patients, assisting them in maintaining their health and promoting a consciousness of wellness. AOM may be especially appropriate for reducing healthcare costs and improving access to care.[6] It is also very safe, resulting in very rare unintended events.[7] Licensed acupuncturists also figure heavily in two state initiatives - in Oregon and Vermont - to limit opioid dependency.

Scope of practice

The scope of practice for AOM varies from state to state. In state statutes, acupuncture is frequently defined as *the stimulation of certain points on or near the surface of the human body by the insertion of needles to prevent or modify the perception of pain, to normalize physiological functions, or to treat certain diseases or dysfunctions of the body.* A number of state statutes also reference the *energetic* aspect of acupuncture by noting its usefulness in controlling and regulating the flow and balance of *qi* in the body or in normalizing energetic physiological function. Other state statutes may define acupuncture by reference to traditional or modern Chinese or Oriental medical concepts or to modern techniques of diagnostic evaluation.

State laws may also specifically authorize acupuncture licensees to employ a wide panoply of AOM therapies such as moxibustion, cupping, Oriental or therapeutic massage, therapeutic exercise, electro-acupuncture, acupressure, dietary recommendations, herbal therapy, injection and laser therapy, ion cord devices, magnets, *qi gong*, and massage. The treatment of animals, the ordering of western diagnostic tests and the use of homeopathy are, with limitations, within the scope of practice in some states.

Referral practices

AOM practitioners receive training in biomedicine, including training on appropriate referrals to medical practitioners. Some states specify the process for referral, but in most cases medical referral is determined by the AOM practitioner. Conversely, patients seeking AOM practitioners are mostly self-referred or referred by word of mouth.

Patients may be referred to professional acupuncturists for a variety of complaints. In 2003, a World Health Organization study, *Acupuncture: Review and Analysis of Reports on Controlled Clinical Trials,* cited over 43 conditions that are treatable with acupuncture. While the list below is not comprehensive, it represents the typical array of complaints for which referral to an acupuncturist is quite appropriate:

- pain management, including musculoskeletal pain, rheumatologic or oncological pain, headache/migraine, and pain palliation at end of life
- neurogenic pain, including neuropathy associated with diabetes mellitus, chemotherapy, and the neuropathy associated with antiretroviral therapy

- women's health, including dysmenorrhea, premenstrual syndrome, amenorrhea, infertility, and pain associated with endometriosis
- nausea and vomiting associated with pregnancy, chemotherapy-induced nausea, and postoperative nausea
- symptoms associated with virally mediated disorders, including HIV, hepatitis B and C viruses, and the pain associated with herpes zoster and herpes simplex I and II
- neurological disorders, including Bell's palsy, cerebral vascular accident, and multiple sclerosis
- routine infections such as colds and flu
- dermatological complaints, such as dermatitis and Psoriasis
- gastrointestinal complaints, including gastralgia, inflammatory bowel disease, irritable bowel syndrome, and gastroesophageal reflux disease
- respiratory complaints, including mild to moderate asthma, shortness of breath, and cough
- urogenital complaints, including dysurea, cystitis, and impotence
- mild to moderate anxiety and depression
- chemical dependency, including abuse of opiates and sympathomimetics
- fatigue secondary to chronic illness, medical treatment, or surgery
- health issues related to aging and imbalance: kidney disease, heart disease, diabetes
- cancer-related issues such as pain and the side-effects of chemotherapy

Third-party payers

Increasingly, acupuncture is covered in many insurance plans, is reimbursed by other third-party payers for certain conditions, and is a part of many employer programs. For the most part, however, payments to individual providers are made by the patient. Indeed, AOM has grown within the US through word of mouth and referrals based on successful treatments and is increasingly recognized as a safe, effective medicine in a wide variety of settings including general health care, integrative medical practices, hospitals, spas and medispa practices, sports and fitness gyms, nonprofit community clinics, and addiction and recovery programs.[7,8,9] There remains, however, a significant disparity in traditional insurance coverage compared to other healthcare professions that have been established in the mainstream world of insurance reimbursement for several decades.

Legislation requiring the coverage of acupuncture services under federal Medicare and the Federal Employee Benefits Program (FEBP) has been introduced in Congress for a number of years. The enactment of this legislation has long enjoyed significant support from within and outside of the profession.[10] Additionally, there is increasing utilization of acupuncture by programs funded by the Department of Defense for active duty personnel and by the Department of Veterans Affairs, and some states have included acupuncture as an Essential Health Benefit under the Affordable Care Act.

Integration Activities

Interest in training that prepares AOM students to work in integrative healthcare settings is high among Council of Colleges of Acupuncture and Oriental Medicine (CCAOM)'s 57 AOM member institutions. A substantial number of these colleges offer internships in off-site clinics where AOM services are provided in local communities. These clinics include a variety of settings, including:

- hospitals
- multi-specialty centers
- research-based centers
- public health clinics
- long- and short-term rehabilitation centers
- family practice clinics
- nursing homes
- outpatient geriatric or assisted living centers for seniors
- drug treatment centers
- HIV/AIDS treatment facilities
- sports medicine clinics
- pediatric
- cancer and other specialty care centers

- clinics addressing specific community group needs, such as women's health and inner city/low income/multi-racial groups

In addition, ACAOM, CCAOM, and NCCAOM actively participate in the work of the Academic Collaborative for Integrative Health (ACIH) whose work, in collaboration with leaders from conventional medical educational institutions, has focused on developing strategies for integrating conventional and IHM education.

Education

Schools and programs

The first acupuncture school in the US was established in 1970.[11] In 1982, the Council of Colleges of Acupuncture and Oriental Medicine (CCAOM) was founded as a nonprofit, voluntary membership association for AOM colleges and programs. Currently, the membership of the CCAOM consists of 57 member schools, all of which have achieved accreditation, or are candidates for accreditation with ACAOM, the national accrediting agency for AOM schools. This membership is spread over 21 states, with major concentrations of colleges in California, New York, and Florida. There is significant diversity in the programs among CCAOM's member schools with representation of the Traditional Chinese Medicine, Japanese, Five Element, Korean, and Vietnamese traditions. The growth of CCAOM since the early years of its founding, when there were fewer than 12 member schools, has mirrored the general growth of the profession. The mission of CCAOM is to support member institutions to deliver educational excellence and quality patient care.

One of the goals supporting the mission of CCAOM is to take a leadership role in acupuncture safety through publication, education, and delivery of a national needle safety course. Pursuant to this goal, the Council's Clean Needle Technique (CNT) course is taught to AOM students to ensure that there is a uniform national course based on best practices for the safe delivery of acupuncture treatment in the interest of public safety. Verification of successful completion of the Council's

CNT course, as evidenced by issuance of a certificate by the Council, is required for a person to receive Diplomate status from NCCAOM.

The Executive Committee of the CCAOM consists of eight officers who are full-time presidents, CEOs, program directors, or other administrators at CCAOM's member schools. The full CCAOM meets twice each year in conferences that provide a forum for member institutions to dialogue about issues of importance to the colleges. In addition, CCAOM's meetings provide an opportunity for its many committees to meet face-to-face and for the membership to benefit from panel discussions, strategic planning, and specialized workshops.

Curriculum content

The entry-level standard of training and education for the profession is the master's degree, which is typically a 3-4 year program that offers either the first-professional master's degree or the first-professional master's level certificate or diploma. Under ACAOM's national accreditation standards, AOM institutions that offer a master's degree in acupuncture must be at least three academic years in length and provide a minimum of 1,905 hours of a professional acupuncture curriculum over a three-year period (typically taken over four calendar years). The curriculum for an acupuncture program must consist of at least 705 didactic hours in Oriental medical theory, diagnosis, and treatment techniques in acupuncture and related studies; 660 hours in clinical training; 450 hours in biomedical clinical sciences; and 90 hours in counseling, communication, ethics and practice management.

A professional Oriental medicine curriculum, which is a minimum of four academic years and includes training in acupuncture and the additional study of Chinese herbal medicine, must consist of at least 2,625 hours leading to a master's degree or master's level certificate or diploma in Oriental medicine. The curriculum for an Oriental medicine program must consist of at least 705 hours in Oriental medical theory, diagnosis, and treatment techniques in acupuncture and related studies; 450 hours in didactic Oriental herbal studies; 870 hours in integrated acupuncture and herbal clinical training; 510 hours in biomedical clinical sciences; and 90 hours in counseling, communication, ethics, and practice management.

For some years, there has been an ongoing dialogue within the profession concerning the amount of academic training a person needs to practice AOM at the entry level and what the most appropriate professional title should be for providers. This dialogue has been formally focused within a doctoral task force established by ACAOM, whose work is discussed in more detail in this chapter under the heading of Accreditation. The average number of academic hours of study and training associated with the entry-level Master's degree among AOM colleges nationally is between 2,700 and 2,900 hours. A significant number of Oriental medicine programs have a curriculum of 3,000 hours or more. The trend over the years has been for an increase in the number of academic hours.

As of August 2016, ACAOM has approved twelve colleges to offer postgraduate clinical doctoral programs (nine have achieved accreditation and three have candidacy status for this degree). The degree for this program, titled Doctorate in Acupuncture & Oriental Medicine (DAOM), provides advanced training in either acupuncture or in Oriental medicine. DAOM programs must provide advanced didactic and clinical training in one or more clinical specialty areas in AOM. Completion of a master's degree or master's level program in acupuncture or in Oriental medicine is a prerequisite for admission into a DAOM program. Such programs must comprise a minimum of 1,200 hours of advanced AOM training in one or more clinical specialty areas at the doctoral level. Within ten years of beginning a DAOM program, a majority of the faculty must possess a doctoral degree, the terminal degree, or its international equivalent in the subject areas in which the faculty teach. Clinical supervisors should have a minimum of five years of documented professional experience as licensed Oriental Medicine practitioners.

After a lengthy development process, ACAOM published competency-based standards for first professional doctorate programs in March 2013 and began accepting substantive change applications for review in June 2013. As of August 2016, six institutions had been granted substantive change approval to begin enrolling students;

however, no programs have been granted candidacy or accreditation status to date.

Faculty and other training information

The accreditation standards for master's level programs as prescribed by ACAOM have general requirements for AOM faculty. Thus, AOM faculty must be academically qualified and numerically sufficient to perform their responsibilities. Additionally, the general education, professional education, teaching experience, and practical professional experience of faculty must be appropriate for the subject area taught. Faculty members must also provide continuing evidence of keeping abreast of developments in the fields and subjects in which they teach. An increasing number of faculty members at AOM institutions now have doctoral degrees. This transition has been facilitated by institutions offering the post-graduate DAOM program.

Accreditation

ACAOM was established in 1982 and is the only national accrediting body recognized by the US Department of Education as a reliable authority for quality education and training in AOM. Its current scope of recognition (effective 9/22/2016) with the USDE is "…the accreditation and pre-accreditation ("Candidacy") throughout the United States of professional non-degree and graduate degree programs, including professional doctoral programs, in the field of acupuncture and/or Oriental medicine, as well as freestanding institutions and colleges of acupuncture and/or Oriental medicine that offer such programs."[12] As both a programmatic and institutional accrediting agency, ACAOM's primary purposes are to establish comprehensive educational and institutional requirements for acupuncture and Oriental medicine programs; to accredit programs and institutions that meet these requirements; and to foster excellence in AOM through the implementation of accreditation standards for AOM educational institutions and programs.

ACAOM was first recognized by the US Department of Education (USDE) in 1988 for the accreditation of master's degree and master's-level acupuncture programs. In 1992, ACAOM was granted an expansion of scope by the USDE to include the accreditation of programs in Oriental medicine, which are programs that include the study of Chinese herbal medicine. In May 2006, the USDE renewed ACAOM's recognition for the maximum five-year period and granted ACAOM's request for an expansion of scope to include candidacy reviews, thus making it possible for nonprofit, freestanding AOM institutions that have achieved candidacy status to establish eligibility for their students to participate in federal Title IV student aid programs; however, USDE's Title IV regulations require for-profit or proprietary AOM institutions to achieve accreditation to establish eligibility for their students to participate in Title IV student aid programs.

Effective July 8, 2011, ACAOM received continued recognition by USDE and ACAOM's authority was expanded to include candidacy and accreditation of DAOM programs, allowing schools that have institutional accreditation from ACAOM to use that status to establish eligibility for Title IV funding for DAOM programs.

ACAOM most recently received continuing recognition from the USDE on September 22, 2016 for the maximum five-year period.

ACAOM currently accredits 57 academic institutions with a total of 63 campuses located in 22 states. ACAOM currently provides institutional or programmatic accreditation or candidacy status to 94 AOM programs. Additional AOM schools that have not achieved candidate or accreditation status with ACAOM do exist in the US.

The recent national trend toward the provision of healthcare services in integrated and conventional settings is reflected in the work of ACAOM's Doctoral Task Force (2003-2012), which was composed of representatives from a broad segment of the AOM profession. The task force drafted outcomes and competency-based accreditation standards for the first-professional entry-level doctorate that take into account the increasing acceptance of AOM practice in conventional and complex medical environments. The task force completed its review of public comments in August of 2012 and forwarded final recommendations to ACAOM, leading to the March 2013 publication of the First Professional Doctorate standards.

Regulation and Certification

Regulatory status

Currently, the right to practice acupuncture by comprehensively trained AOM practitioners exists in 46 states and in the District of Columbia as described in each state's practice act. The right to practice may be designated by licensure, certification, or registration under the applicable state law. Licensure is the most common form of authorization to practice. In states without regulation, practitioners typically practice subject to potential oversight from a medical board. Other states may limit practice specifically to designated medical providers or AOM practitioners who are medically supervised. In the few remaining unregulated states, activities to obtain full licensure status for acupuncturists are underway.

In most states with regulation, professional acupuncturists have independent status, although there remain a few states where practitioners must have supervision, prior referral, or initial diagnosis by a conventional medical doctor. The recent statutory trend, however, is in favor of more professional independence by AOM providers. The most common designation for comprehensively-trained practitioners is Licensed Acupuncturist (LAc), although in a few states they may be designated by statute as Acupuncturist Physicians (AP-Florida) or Doctors of Oriental Medicine (DOM-New Mexico and Nevada), but these doctoral designations are licensure titles conferred by the state and do not reflect earned academic degrees at the doctoral level. The State of Washington now specifies Eastern Asian Medicine Practitioner (EAMP) although the revised statute allows the continued use of Licensed Acupuncturist (LAc).

The state statutes regulating acupuncture are not uniform. In some states there are very detailed statutes and regulations, but in others there may be only a few paragraphs concerning the practice of acupuncture. The administrative structure for the regulation of acupuncture in the states also varies considerably. The most common structure is for the profession to be regulated by an independent board composed of professional acupuncturists or by a state medical board with the assistance of an advisory acupuncture board or committee.

Other administrative arrangements include regulation by a joint board comprised of diverse healthcare professionals (both conventional and IHM), by the board of another IHM profession, or by a larger administrative division within a state department—all typically with the assistance of an acupuncture advisory body. This diversity reflects political and budgetary realities as each state tailors its law to meet local needs.

All states regulating the practice of acupuncture require or accept passage of NCCAOM certification examinations or, in the case of 23 states, require full NCCAOM certification as either a Diplomate of Acupuncture (Dipl. Ac. [NCCAOM]) or Diplomate of Oriental Medicine (Dipl. O.M. [NCCAOM]). Currently, the only state not using the NCCAOM certification exams for licensure is California. California provides its own licensing exam. Figure 1.1 illustrates those states that use NCCAOM examinations or require full certification by NCCAOM. In June of 2016 the California Acupuncture Board (CAB) passed a motion to recommend a regulatory change to have the CAB use the NCCAOM examinations for the licensing of acupuncturists in that state no earlier than 2019.

In general, national AOM organizations have been supportive of the adoption of state acupuncture practice acts that incorporate reference to national standards of education, training, and certification in the field. Adherence to such standards, as administered by ACAOM in the field of accreditation and by NCCAOM for certification, assures a high level of practitioner competence and a degree of uniformity that facilitates portability among the various states in the recognition of practitioner credentials.

Figure 1.1

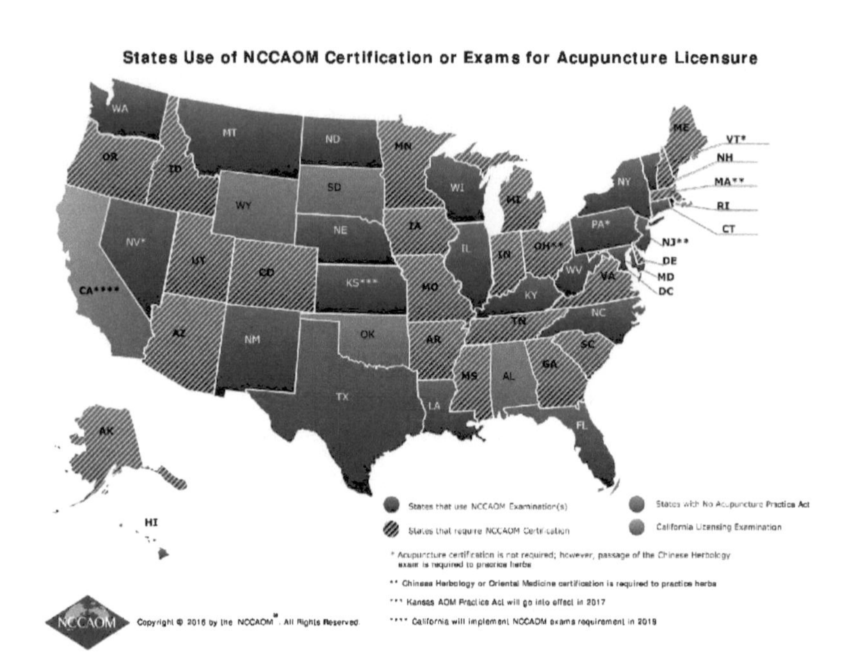

Examinations and certifications

Graduates of ACAOM accredited or candidate institutions are qualified to take the national certification exams offered by NCCAOM. Established in 1982, the mission of NCCAOM is to assure the safety and well-being of the public and to advance the professional practice of acupuncture and Oriental medicine by establishing and promoting national evidence-based standards of competence and credentialing. NCCAOM is a member of the Institute for Credentialing Excellence (ICE) and its certification programs in Acupuncture, Chinese Herbology, and Oriental Medicine are accredited by the National Commission for Certifying Agencies (NCCA), a separate independent commission of ICE. Accreditation by NCCA ICE represents the highest voluntary certification standards in the US. Passage of two or more of NCCAOM's national examinations are a requirement of licensure recognized in 44 states and the District of Columbia. Although California has its own exam, movement is underway to require all four NCCAOM certification exams in the near future. Candidates who pass NCCAOM's required certification exams in Acupuncture, Chinese Herbology, and Oriental Medicine are awarded the designation NCCAOM Diplomate appropriate to the certification achieved: Dipl. Ac. (NCCAOM)®, Dipl. O.M. (NCCAOM) ®, and Dipl. C.H. (NCCAOM)®.

The first NCCAOM comprehensive written examination in acupuncture was administered in 1985 and was developed over a three-year period with the assistance of leading acupuncturists throughout the US. In 1989, NCCAOM added a practical examination of point location skills as a component of its acupuncture examination. A clean needle technique exam was added to the certification requirements for the acupuncture written exam in 1991 and merged into the acupuncture exam in 1998. The NCCAOM administered the first national examination in Chinese Herbology in 1995 and in 2000 offered its first written examination in Asian Bodywork Therapy (Note: Although the NCCAOM still recertifies current Diplomates of Asian Bodywork Therapy, it has not offered any new certifications in this program since 2012). In 2003, NCCAOM began to offer an umbrella certification in Oriental medicine to applicants who demonstrated competence in both acupuncture and Chinese herbology, as well as entry-level competency in biomedicine.

Since its inception, the NCCAOM has issued more than 25,000 certificates in acupuncture, Oriental Medicine, Chinese Herbology, and Asian Bodywork Therapy and reports the existence of more than 17,000 active Diplomates worldwide in current practice. Figure 1.2 shows the number of actively certified (i.e. newly certified or recertified) NCCAOM Diplomates per state. The NCCAOM requires all active Diplomates to recertify every four years by submitting documentation of 60 professional development activity points.

Figure 1.2

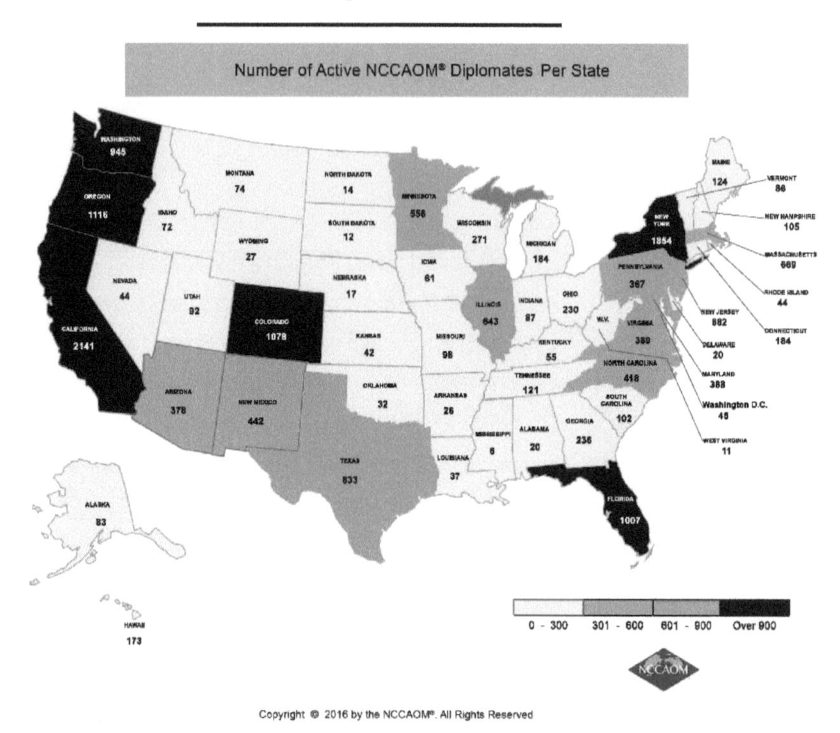

Research

Research in the US on the efficacy of AOM increased after 1996 when the FDA reclassified acupuncture needles from experimental devices (Class III) to devices for which performance standards exist (Class II). This provided reasonable assurance that acupuncture needles would be safe, thus making research easier to conduct. Subsequent investigations over the last decade have found that acupuncture needles had few side effects, making them somewhat safer than certain conventional western medical treatments for many diseases. Research in AOM has increased dramatically over the past decade within the US, as well as in Asia, Europe, and South America. Specifically, a recent review found exponential growth in publications over the past two decades, with a mean annual growth rate of 10.7%.[13]

Among the more prominent scientific endorsements of the efficacy of acupuncture is that of the National Institutes of Health, which concluded in a 1997 Consensus Statement that:

> *...promising results have emerged, for example, showing efficacy of acupuncture in adult postoperative and chemotherapy nausea and vomiting and in postoperative dental pain. There are other situations such as addiction, stroke rehabilitation, headache, menstrual cramps, tennis elbow, fibromyalgia, myofascial pain, osteoarthritis, low back pain, carpal tunnel syndrome, and asthma, in which acupuncture may be useful as an adjunct treatment or an acceptable alternative or be included in a comprehensive management program. Further research is likely to uncover additional areas where acupuncture interventions will be useful.*[14]

Within the AOM profession, the most prominent research organization in the US is the Society for Acupuncture Research (SAR), which was formally established in 1993 and whose mission is to promote, advance, and disseminate scientific inquiry into Oriental medicine systems, which include acupuncture, herbal therapy and other modalities. SAR values quantitative and qualitative research addressing clinical efficacy, physiological mechanisms, patterns of use, and theoretical foundations. The organization sponsors biennial symposia on research methodologies and welcomes individual affiliates including researchers, educators, students, acupuncturists, healthcare practitioners, and members of the public, as well as institutional affiliates including schools, vendors, and other organizations to participate in these events.

In November 2007, SAR held an international symposium aimed at presenting and discussing the progress made in acupuncture research during the decade following the 1997 NIH Consensus Development Conference. The symposium presentations, as well as their summaries are published in the *Journal of Alternative and Complementary Medicine.*[15-17] These reports unequivocally show that the field of acupuncture research has significantly expanded and matured since 1997. Phase II/III sham-controlled trials have been successfully completed and a broad range of basic research studies have identified numerous biochemical and physiological correlates of acupuncture. However, SAR has also identified intriguing paradoxes emerging from the symposium and summaries that touch on some disconnects, including those between clinical and basic acupuncture research. A discussion of these issues was published in 2011.[18] Subsequent SAR White Papers explore additional overarching topics related to acupuncture research ("Manual and Electrical Needle Stimulation in Acupuncture Research: Pitfalls and Challenges of Heterogeneity," and "Unanticipated Insights into Biomedicine from the Study of Acupuncture"), with a fourth white paper currently being written with the topic: the role of acupuncture and Chinese medicine in personalized/precision medicine initiatives within our health care system. These white papers are available for public access: http://www.acupunctureresearch.org/white-papers.

In 2011, CCAOM established a pilot research grant program to encourage collaborative research proposals and improve the research capacity of its member colleges. This initiative is implemented annually through a call for research proposals from CCAOM's member colleges. To date, grant awards have been approved for the following research projects:

- Relationship between biomarkers of aging and qigong practice: a case control study
- Premenstrual syndrome research survey
- The feasibility and acceptability of adjuvant acupuncture to lidocaine as a treatment for provoked, localized vulvodynia: a pilot trial.
- Acupuncture for Chemotherapy-Induced Peripheral Neuropathy- a Case Series
- Scientific Worldview Differences Between Students in Biomedical and East Asian Medical Degree Programs and the Impact of an Educational Exchange Program

SAR holds biennial conferences and serves as a venue for timely reporting of the latest acupuncture research highlights destined for high impact research journals. These conferences are also an excellent forum for strengthening research curricula in CCAOM's member schools. The most recent SAR conference, in 2015, focused on increasing the scope of acupuncture research and shared one whole research day with two other annual conferences: the Society for Integrative Oncology and the Fascia Research Society. This 'shared-day,' which attracted more than 500 researchers, practitioners, students, and funding agency personnel from many different countries, was jointly sponsored by SAR, The Osher Foundation, Society for Integrative Oncology, and Fascia Research Society. SAR's next international conference, planned for April 2017 will be co-sponsored by the Department of Anesthesia at Stanford University and will be held in San Francisco, CA.

Challenges and Opportunities

Key challenges 2016 – 2020

- need for greater consensus concerning what primary care competencies are appropriate for LAc's
- need for greater public awareness of the skill limitations of other healthcare providers who practice acupuncture with minimal training, without having completed a full three- or four-year AOM curriculum at an accredited AOM institution or program.
- need for greater reciprocity among states in recognizing the credentials of AOM providers
- need for an appropriate response to increasing professional and public expectations for integrated health care
- regulatory uncertainty concerning FDA restrictions on the use of Chinese herbs
- need for expanded reimbursement for AOM treatments by HMOs, Medicare, and third-party payers, and for identification of practitioner qualifications/standards for obtaining such reimbursement
- need for greater emphasis on AOM research, training of AOM researchers, and increased funding for AOM research

- need to strengthen and coordinate national representation and AOM profession's voice for federal policy issues
- need for greater public awareness and acceptance of AOM with associated increase in viable professional practice opportunities for a greater number of AOM practitioners many of whom graduate with substantial student loan debt
- need for greater awareness of the qualifications of licensed acupuncturists in the provider credentialing process at hospitals and major medical centers
- need for more comprehensive and coordinated collection of data concerning the profession

Key opportunities 2016 – 2020

- work of ACAOM to implement the first-professional doctorate (FPD) and review/revise the existing Master Standards consistent with the new competency based standards
- increase in number of AOM schools offering postgraduate clinical education through the Doctorate of Acupuncture and Oriental Medicine programs
- work of ACIH focusing on interdisciplinary IHM and conventional healthcare education
- growing commitment within the profession for greater collaboration within the field, including joint projects among the national and state organizations
- increased marketing activities to develop public awareness of the benefits, safety, and cost-effectiveness of AOM
- success and visibility of emergency acupuncture services provided in aftermath of national disasters which are a precedent for similar opportunities in future emergency situations and for greater coordination within the profession for local emergency response
- expansion of education and accreditation to include advanced or specialized training in AOM areas including bodywork (*tui na* and *shiatsu)*, nutrition, meditation, and exercise

Resources

Organizations and websites

- Accreditation Commission for Acupuncture and Oriental Medicine (ACAOM) www.acaom.org
- Acupuncture Now Foundation (ANF) https://acupuncturenowfoundation.org
- Acupuncturists Without Borders (AWB) www.acuwithoutborders.org
- American Association of Acupuncture and Oriental Medicine (AAAOM) www.aaaomonline.org
- American Organization for Bodywork Therapies of Asia (AOBTA) www.aobta.org
- American Society of Acupuncturists (ASA) www.asacu.com
- Council of Colleges of Acupuncture and Oriental Medicine (CCAOM) www.ccaom.org
- National Acupuncture Detoxification Association (NADA) www.acudetox.com
- National Certification Commission for Acupuncture and Oriental Medicine (NCCAOM®) www.nccaom.org
- Society for Acupuncture Research (SAR) www.acupunctureresearch.org

Bibliography:

Bensky D, Barolet R, Ellis A, Scheid V. *Chinese Herbal Medicine Formulas and Strategies*. 2nd ed. Seattle, WA: Eastland Press; 2009.

Bensky D, Clavey S, Stroger E. *Chinese Herbal Medicine Materia Medica*, 3rd ed. Seattle, WA: Eastland Press; 2004.

Chen J, Chen T. *Chinese Herbal Formulas and Applications: Pharmacological Effects and Clinical Research*. City of Industry, CA: Art of Medicine Press; 2009.

Chen J, Chen T. *Chinese Herbal Medicine and Pharmacology*. City of Industry, CA: Art of Medicine Press; 2003.

Cheng X, ed. *Chinese Acupuncture and Moxibustion*. Beijing, China: Foreign Languages Press; 1987.

Deadman P, Al-Khafaji A. *A Manual of Acupuncture*. Sussex, England: Journal of Chinese Medicine Publications; 1998.

Huang Ti Nei Ching Su Wen. Veith I, trans. Berkeley, CA: University of California Press; 1949.

Maciocia G. *The Foundations of Chinese Medicine*. Philadelphia, PA: Churchill Livingstone; 1989.

Unschuld P, trans. *Nan Ching*. Berkeley, CA: University of California Press; 1986.

Unschuld P. *Huang Di Nei Jing Su Wen: Nature, Knowledge, Imagery in an Ancient Chinese Medical Text*. Berkeley, CA: University of California Press; 2003.

Unschuld P. *Medicine in China: A History of Ideas*. Berkeley, CA: University of California Press; 1988.

Wiseman N, Ye F. *A Practical Dictionary of Chinese Medicine*. Brookline, MA: Paradigm Publications; 1998.

Zmiewski P, Wiseman N, Ellis A, eds. *Fundamentals of Chinese Medicine*. Brookline, MA: Paradigm Publications; 1985.

Citations

1. Reston J. Now let me tell you about my appendectomy in Peking. *New York Times*. July 26, 1971
2. Wang Z. Foundations of Oriental Medicine, Biomedicine, Acupuncture with Point Location, and Chinese Herbology Job Analysis Report 2013; NCCAOM® and Schroeder Measurement Technologies, Inc.; http://www.nccaom.org/job-analysis/, accessed 12/19/16
3. Ward-Cook, K; Descriptive Demographic and Clinical Practice Profile of Acupuncturists: An Executive Summary from the NCCAOM® 2013 Job Analysis Survey; NCCAOM® 2014; http://www.nccaom.org/job-task-analysis-jta-informational-page/, accessed 12/19/16
4. ACAOM Reports Enrollment and Graduation Numbers. http://acaom.org/
5. New Independent SOC for Acupuncturists proposed by the BLS for 2018. http://www.nccaom.org/resource-center/bls-timeline/
6. Jabbour M, Sapko MT, Miller DW, et al. Economic evaluation in acupuncture: past and future, *Am Acupuncturist*. 2009; Fall(49):11.

7. Xu S, Want L, Cooper E, et al. Adverse events of acupuncture: a systematic review of case reports. *Evid Based Complement Alternat Med*. 2013, Vol. 2013: 1-15.
8. Witt CM, Jena S, Brinkhaus B, et al. Acupuncture in patients with osteoarthritis of the knee or hip: a randomized, controlled trial with an additional nonrandomized arm. *Arthritis Rheum*. 2006, Nov; 54(11):3485-93.
9. Cherkin DC, Sherman KJ, Avins AL, et al. A randomized trial comparing acupuncture, simulated acupuncture, and usual care for chronic low back pain. *Arch Intern Med*, 2009;169(9):858-866.
10. 4839, 114 Cong., 2015 (enacted). Print.
11. Cohn S. Acupuncture, 1965-85: Birth of a new organized profession in the United States. *Am Acupuncturist*. 2010; 54(Winter): 12-15.
12. Renewal of Recognition Letter, US Dept. of Education. 2016 https://www.insidehighered.com/sites/default/server_files/files/ACAOM S2016 Senior Department Official s (SDO) Decision Letter.PDF
13. Ma Y, Dong M, Zhou K, Mita C, Liu J, Wayne PM. Publication Trends in Acupuncture Research: A 20-Year Bibliometric Analysis Based on PubMed. *PLoS ONE*, 2016;11(12): e0168123. doi:10.1371/journal.pone.0168123
14. Acupuncture. NIH Consensus Statement 1997 Nov 3-5; 15(5):1-34.
15. MacPherson H, Nahin R, Paterson C, Cassidy CM, Lewith GT, Hammerschlag R. Developments in acupuncture research: big-picture perspectives from the leading edge. *JACM*. 2008;14(7):883-7.
16. Park J, Linde K, Manheimer E, Molsberger A, Sherman K, Smith C, Sung J, Vickers A, Schnyer R. The status and future of acupuncture clinical research. *JACM*. 2008;14(7):871-81.
17. Napadow V, Ahn A, Longhurst J, Lao L, Stener-Victorin E, Harris R, Langevin HM. The status and future of acupuncture mechanism research. *JACM*. 2008 Sep;14(7):861-9. PMID: 18803495

18. Langevin H, Wayne P, MacPherson H, et al. Paradoxes in acupuncture research: strategies for moving forward, *Evid Based Complement Alternat Med*, 2011, doi:10.1155/2011/180805.

Chiropractic Health Care

Third Edition (2017) Editors: Joseph Brimhall, DC;
David O'Bryon, JD, CAE

Second Edition (2013) Editors: Michael Wiles, DC, MEd, MS;
David O'Bryon, JD, CAE; Joseph Brimhall, DC

First Edition (2009) Authors: Reed Phillips, DC, PhD;
Michael Wiles, DC, MEd, MS; David O'Bryon, JD, CAE

Partner Organization: Association of Chiropractic Colleges

About the Authors/Editors: Phillips is former President of the Southern California University of Health Sciences, past Vice President of the Foundation for Chiropractic Education and Research, and a past President of both the Association of Chiropractic Colleges and the Council on Chiropractic Education. Wiles is Dean at the Keiser University College of Chiropractic and past Co-chair of the Academic Collaborative for Integrative Health Education Working Group. O'Bryon is President of the Association of Chiropractic Colleges, Chair of ACIH, past President of the Federation of Associations of Schools of the Allied Health Professions and serves on the Secretariat Board of the National Association of Independent Colleges and Universities. Brimhall is President of the University of Western States, Chair of the Northwest Commission on Colleges and Universities, Chair of the Oregon Collaborative for Integrative Medicine, a past member of the ACIH Executive Committee, past President of the Councils on Chiropractic Education International, and a former Board President and Commission Chairman of the Council on Chiropractic Education.

Philosophy, Mission, Goals

Chiropractic is a healthcare discipline that emphasizes the inherent power of the body to heal itself. A Doctor of Chiropractic practicing primary health care is competent and qualified to provide independent, quality, patient-focused care to individuals of all ages and genders by: 1) providing direct access, portal-of-entry care that does not require a referral from another source; 2) establishing a partnership relationship with continuity of care for each individual patient; 3) evaluating a patient and independently establishing a diagnosis or diagnoses; and, 4) managing the patient's health care and integrating healthcare services including treatment, recommendations for self-care, referral, and/or co-management.[1]

The above definition is taken from January 2013 Accreditation Standards from the Council on Chiropractic Education (CCE), the chiropractic accrediting agency recognized by the United States Department of Education. This definition is used by Doctor of Chiropractic degree programs in establishing educational curricula and clinical competencies that prepare graduates for professional chiropractic practice. Preferences and practices vary in how the profession describes itself.

A Doctor of Chiropractic may, depending on jurisdictional definitions and individual preference, utilize any of a range of titles: chiropractor, doctor of chiropractic, doctor of chiropractic medicine, or chiropractic physician. Some will only use "chiropractic" to describe the practice while others prefer "chiropractic medicine" or "chiropractic health care." This range of usages is also found in statutory and regulatory language. In some jurisdictions, laws establish the terms and titles that are used. The Department of Labor description includes the terms "chiropractic physician" and "chiropractic medicine."

Inside the Association of Chiropractic Colleges (ACC), individual member institutions may also choose diverse designations. For this reason we will intersperse mixed use of these terms in this chapter. We urge practitioners from other disciplines to engage the members of the chiropractic profession on their legally established titles and preferences. This may provide a useful way to improve understanding and collaboration.

Early chiropractic concepts of health and disease were compatible with vitalistic philosophies of the late 19th century. Living cells were conceived as having an inborn intelligent component that was responsible for the maintenance of life. This phenomenon is analogous to homeostasis. Some homeostatic control mechanisms function as a result of coordination of sensory input and motor output through the central nervous system. The intimacy of body structure (particularly the spine) and the nervous system, and hence body function, led the early chiropractors to develop a philosophy of care that, in essence, posited that a normally functioning nervous system, in the presence of structural normality, should lead to normal health. This vitalistic approach toward health and disease was similar to that proposed by the early osteopathic profession.

History of the profession

Chiropractic was founded by Daniel David Palmer in Davenport, Iowa in 1895. Palmer was originally a schoolteacher who developed an interest in magnetism and what was, at that time, popularly called magnetic healing. Having observed that a patient's deafness had apparently occurred following a traumatic experience resulting in a protuberance on his spine, Palmer reasoned that reduction of this protuberance might affect the patient's hearing. Evidently, following a manipulation of the spine, the patient's hearing improved. This led to the formulation of a theory relating spinal alignment to states of health and disease, which was similar to that espoused by Andrew Taylor Still, the founder of osteopathy, in 1874 in neighboring Missouri. In 1975, the National Institutes of Health sponsored an interdisciplinary conference called "The Research Status of Spinal Manipulative Therapy" in Bethesda, Maryland. This conference brought together leaders in the field of spinal manipulation from the chiropractic, medical, and osteopathic professions from around the world. It marked a significant point in the evolution of chiropractic education and practice, and important progress has been made since that time toward the integration of chiropractic education, science, research and practice into mainstream systems of health care and professional education. While this process is far from complete, the chiropractic profession of today can be proud of the great advances in education and research of the past few decades.

Chiropractic spread first to Canada, then to the United Kingdom, and finally throughout the world from these simple roots in Davenport, Iowa. Educational institutions offer chiropractic degrees in Europe, Africa, Australia, Asia, and North and South America, with new programs currently under development. The World Health

Organization has published guidelines on basic training and safety in chiropractic education and practice.[2]

Characteristics and Data

The chiropractic medical profession is growing rapidly throughout the world. There are approximately 77,000 chiropractors in the US and about another 22,750 in the rest of the world.[3,4] Doctors of chiropractic are regulated as members of a licensed healthcare profession in all jurisdictions of the United States and Canada, in Mexico, and in about 40 other countries. In roughly 60 additional countries, chiropractors practice in an unregulated fashion, and many of these countries are currently in the process of officially recognizing and regulating the practice of chiropractic health care. In countries where chiropractic medicine is regulated, many of the chiropractic degree programs and colleges are located within university settings.

Income data vary for chiropractors, depending on the country and source of the data. Most sources suggest that full-time chiropractic physicians earn an income comparable with other health professionals. In a recent annual salary and expense survey, chiropractors reported an average total compensation of $138,000.[5]

Clinical Care

Approach to patient care

Today, the traditional vitalistic philosophy has evolved to a patient care approach that recognizes and honors the body's own innate mechanisms for adaptation and homeostasis. Treatment is designed to optimize and support the body's natural self-healing and intrinsic regulatory systems.

In practical terms, this can be translated into an approach to care that seeks to relieve symptoms, restore joint motion, enhance posture and balance, provide necessary support, strengthen muscles, improve flexibility and facilitate coordination to optimize body function. A patient may seek help for lower back pain, for example, and this approach to care will address not only the local factors associated with the back pain, but also the structural adaptations that may have led to it in the first place, such as weak muscles, obesity, detrimental postures, and other factors. Diet, activities of daily living, lifestyle behaviors and stress reduction may also be addressed.

Such an approach can be considered both patient-centered and holistic. Chiropractic physicians generally approach patient care in a manner that is similar to conventional medical doctors. Procedurally, the patient is interviewed, a detailed health history is obtained, an examination is performed including any necessary specialized tests, results are compiled and reviewed, and a working diagnosis is formulated. Then, a comprehensive management plan is constructed and recommended, and the patient is afforded an opportunity to provide informed consent before initiating any treatment. Clinical progress is monitored, and the patient is discharged from active care when the appropriate outcomes have been achieved.

Chiropractic physicians establish a standard medical history and are particularly concerned with the identification of factors or conditions that may require either referral to, or co-treatment with, other healthcare providers. Patients commonly seek chiropractic services for complaints related to the musculoskeletal system; accordingly, musculoskeletal diagnoses are common in chiropractic practices.

Doctors of chiropractic focus not only on the structural component of a patient's complaint, but also evaluate the overall health of the patient. For example, a lower back complaint may be related locally to a sacroiliac joint dysfunction, and generally to poor posture and obesity, all of which must be taken into consideration in the comprehensive treatment and management plan.

Clinical management plans are developed to provide symptom relief and achieve problem resolution, and to optimize whole-person health and function. Typically, treatment involves manual therapy, often including spinal manipulation, and other forms of treatment are commonly used as well. These may include exercise, stretching, rehabilitative measures, physical therapeutics (such as electrotherapy, hydrotherapy, or ultra-sound), diet and nutritional counseling, lifestyle advice, and recommendations for stress reduction. Manual therapy includes a continuum of treatment methods ranging from stretching and sustained pressure techniques to specific joint manipulation, often referred to as chiropractic "adjustments". Some techniques employ articulated treatment tables, and some use mechanical devices that move, stretch or position body structures. There is a very wide spectrum of techniques and methods under the umbrella of manual therapies, and the one most commonly associated with chiropractic treatment is the chiropractic manipulation or "adjustment" typically delivered by hand with a very specific and quick, gentle thrust.

The purpose of the chiropractic adjustment is to improve joint motion and function. The theoretical model is a dysfunctional articular lesion, sometimes traditionally or historically referred to as a subluxation. Joint dysfunction may be associated with a correctible segmental mechanical phenomenon characterized by asymmetry, restriction of motion, tissue texture abnormalities, and tenderness to palpation. Neurologically, spinal joint dysfunction, or subluxation, has been associated with segmental facilitation, and with trophic changes. This "manipulable lesion," as some have called it, is widely known to practitioners of manual medicine and spinal manipulation but is not generally known or appreciated by practitioners outside of those fields. Early chiropractic theories of a "bone out of place pinching a nerve" have plagued the profession for decades and have hindered interprofessional dialogue with regard to this lesion and phenomenon. A new generation of chiropractors and chiropractic educators are making strides toward standardization of terminology, the elucidation of the properties of segmental joint dysfunction, and the biomechanical effects of spinal manipulative therapy.

The majority of patients consult a chiropractor for complaints directly related to back pain and other musculoskeletal complaints. Patients also seek care from chiropractic physicians for headaches, digestive problems, hypertension, fatigue, and many other conditions that are typically seen in a primary healthcare setting.

Scope of practice

The scope of practice for doctors of chiropractic is established by jurisdictional laws and regulations, similar to all other licensed professions. Three features of legislation and practice are common in all jurisdictions that regulate chiropractic health care:

1. primary access, in which doctors of chiropractic accept patients directly without the requirement of referral from any other source
2. authority and obligation to establish a diagnosis prior to the initiation of treatment, which includes the authority to perform examinations and to order necessary diagnostic studies
3. authority to manage patient care by directly providing treatment, referring to another provider for additional care, and recommending lifestyle changes to facilitate health and wellness

A number of jurisdictions also allow chiropractic physicians to perform minor surgery, practice obstetrics, and prescribe medication. The scope of practice for chiropractic medicine varies widely among jurisdictions, and generally includes the diagnosis and treatment of human conditions. All manual treatment methods and the use of adjunctive approaches such as nutritional therapy, counseling, and physical therapeutics are also commonly included in chiropractic regulation.

Referral practices

Chiropractic students are taught the importance of proper and appropriate referral to other healthcare providers. Chiropractic medical education emphasizes the identification of conditions that may require specialized care and incorporates the protocols and processes for referring patients to practitioners that can provide such care. For example, since back pain can be due to a variety of diseases and medical conditions, some of which require urgent referral (such as cauda equina syndrome), doctors of chiropractic are trained to detect and manage patients with such conditions, including any necessary referral. Practicing chiropractic physicians seek to develop referral relationships with conventional medical physicians, osteopathic physicians, and naturopathic physicians, as well as specialists such as orthopedic surgeons, rheumatologists, neurologists, and gynecologists. Patient needs may also require referral to other practitioners, such as massage therapists, acupuncturists, physical therapists, occupational therapists, clinical psychologists, or other licensed and qualified healthcare practitioners.

Many doctors of chiropractic accept patients upon the referral of other healthcare providers. Specifically, patients with complaints related to the musculoskeletal system, such as back pain, neck pain, mechanical headaches, sports injuries, repetitive strain injuries, motor vehicle accident injuries, work-related musculoskeletal injuries, overuse syndromes, and other similar conditions, are frequently referred by other practitioners to chiropractic physicians for evaluation and treatment. Chiropractic physicians establish mutual referral networks with other physicians and healthcare providers. Chiropractors typically also develop a referral network with other chiropractic physicians, particularly those specializing in certain conditions. Chiropractic physicians are trained to establish and maintain records of patient care, and this information is routinely provided to other health practitioners that are involved in the care of a specific patient.

Third-party payers

Insurance coverage and third-party payment for chiropractic medical services vary widely from jurisdiction to jurisdiction. Chiropractic medicine is widely covered through private insurance plans in most countries, particularly in the US. Most government sponsored workers' compensation plans cover chiropractic services. There is an increasing tendency for chiropractic medicine to be included in wellness programs or other similar employer-sponsored health plans.

Integration Activities

All educational institutions offering chiropractic education have student clinics where the underserved or uninsured may receive chiropractic care at little or no cost. A growing number of community clinics include chiropractic medical services alongside the services of medical doctors and other healthcare services. Chiropractic physicians are increasingly involved in larger clinics and hospitals, and a growing number of hospitals grant privileges for chiropractors to treat patients on an outpatient basis and to use diagnostic facilities.

There are many other clinical settings where doctors of chiropractic are part of a multidisciplinary team. An outstanding example is the doctors of chiropractic who serve in the Walter Reed National Military Medical Center. Their presence at this facility is the outcome of ongoing negotiations with the US Department of Defense to employ Doctors of Chiropractic in military healthcare facilities as directed by Congress and the President of the United States.

Following recent legislation, chiropractic medicine is gradually becoming available throughout the military healthcare system. Experience to date has shown that chiropractic physicians in military healthcare facilities quickly become integral and valued members of the healthcare team.

With Presidential insistence, chiropractic medicine has also been introduced into the US Veterans Administration healthcare system. A three-year dialogue with a Federal Advisory Committee consisting of medical and osteopathic physicians, a physical therapist, a physician's assistant, and several doctors of chiropractic resulted in a list of 68 recommendations to the Secretary of the Veteran's Administration on the implementation process. Of the 68 recommendations, 67 were unanimously agreed upon. This program is slowly expanding throughout the entire VA system and current laws mandate the inclusion of chiropractic medical services at all VA facilities within the next few years. Several chiropractic educational institutions now have agreements with their local VA hospitals allowing clinical rotations of senior chiropractic students and interns, and facilitating interaction of conventional medical and chiropractic students, interns, and residents. In 2016 the VA established a formal chiropractic residency program which is accredited by an agency recognized by the US Department of Education.

Chiropractic has become a highly accepted form of treatment in the world of sports. Doctors of Chiropractic (DCs) are utilized by all 32 National Football League teams, and most other major professional teams in all sports employ a team of chiropractic physicians.[6] Many college and university teams engage the services of a doctor of chiropractic as well. Doctors of Chiropractic provide care in many international events such as wrestling, track and field, swimming, and other sports, and are increasingly participating on multidisciplinary healthcare teams at the Olympics and in professional sports. For example, chiropractic physicians are on staff at the US Olympic Training Center in Colorado Springs, Colorado.

As the world of integrative medicine and complementary and integrative health care continues to grow, doctors of chiropractic medicine are at the forefront in many areas and are providing leadership to help establish standards and achieve cultural recognition of the benefits of complementary and integrative health care. Several educational institutions housing accredited doctor of chiropractic degree programs have incorporated other academic programs in acupuncture and Oriental medicine within their structure, at least two have added a program in naturopathic medicine, and many teach the principles and application of homeopathic formularies, clinical nutrition, functional medicine, life-style counseling, mind-body medicine, yoga, and massage therapy.

Over the last decade of increased integration activity, individual Doctors of Chiropractic and leaders of chiropractic programs have engaged various interprofessional and inter-institutional relationships with integrative medical programs that are among the more than 70 member programs of the Academic Consortium for Integrative Medicine & Health. Among these are programs at Yale University, Harvard University, the University of Minnesota, and Oregon Health and Sciences University (OHSU). In the latter instance, the president of University of Western States serves as the chair of the Oregon Collaborative for Integrative Medicine (OCIM), where his colleagues include academic leaders at Oregon Health & Science University, Pacific University, Oregon College of Oriental Medicine and National University of Natural Medicine, all located in Portland, Oregon. The direction of growth and development is toward a more integrated practice among all healthcare disciplines.

Education

Schools and programs

Chiropractic colleges in the US grant the Doctor of Chiropractic degree after a course of study which consists of a minimum of 4,200 classroom hours, typically delivered over a 4–5 academic year program. There are sixteen educational institutions at nineteen campuses in the US and two in Canada that offer Doctor of Chiropractic degree programs. These include individual chiropractic colleges, colleges within private universities, and a college within a public university system. Typically, chiropractic students are in their mid-twenties, with about 75% having completed undergraduate degrees before entering a chiropractic college.

At least 90 semester hours (three years) in undergraduate studies in the biological, physical and social sciences are required prior to admission to chiropractic medical programs in the United States and Canada. Outside North America, chiropractic degree programs are designed to be educationally similar to other direct-access, first-professional healthcare programs. After successfully completing the program of studies, the graduate earns a Doctor of Chiropractic (DC) degree in the US.

Outside of the US, graduates of chiropractic programs may receive other academic designations that are consistent with local jurisdictional requirements and customs. The most common of these credentials are the Bachelor of Science and Master of Science degrees in Chiropractic. In the United Kingdom, a Bachelor of Chiropractic (BChir) is awarded, analogous to the Bachelor of Medicine (MB or BMed) degree.

The Association of Chiropractic Colleges (ACC) represents all accredited colleges in the US and several others from around the world. Programs outside the US are most commonly affiliated with public universities. Accredited chiropractic degree programs exist in Canada (2), United Kingdom (2), Denmark, France, South Africa (2), Australia (3), New Zealand, Switzerland, Korea, Malaysia and Japan. Additional programs in England and Spain (2) hold accreditation candidate status. Several other chiropractic programs around the world are pursuing accreditation, and some are new or under new development including Mexico (2), and Brazil (2).

Curriculum content

The chiropractic curriculum typically includes courses in:

- Anatomy
- Biochemistry
- Physiology
- Microbiology and Immunology
- Pathology
- Public Health
- Clinical Skills (including history and physical examination)
- Clinical and Laboratory Diagnosis
- Clinical Sciences (including the study of cardiopulmonary, gastrointestinal and genitourinary disorders; dermatology; ophthalmology; otolaryngology)
- Gynecology and Obstetrics
- Pediatrics
- Geriatrics
- Diagnostic Imaging (procedures and interpretation)
- Psychology and Abnormal Psychology
- Nutrition and Clinical Nutrition
- Biomechanics
- Orthopedics
- Neurology
- Emergency Procedures and First-Aid
- Spinal Analysis
- Principles and Practice of Chiropractic
- Clinical Reasoning and Decision Making
- Chiropractic Manual Therapy and Adjustive Procedures
- Research Methods and Statistics
- Professional Practice Ethics and Office Management

There are many opportunities for postgraduate study in chiropractic. A number of full-time residency programs exist, of which the most popular and ubiquitous is diagnostic imaging (a three-year, full-time residency). A full-time residency program in chiropractic geriatrics was initiated at Northwestern Health Sciences University, and National University of Health Sciences offers three-year residency programs in family practice and research. Los Angeles College of Chiropractic (Southern California University of Health Sciences) also offers residencies in Chiropractic Sports Medicine and Primary Spine Care Practitioner.

Other institutions are offering additional degree programs. Numerous certification programs exist in a variety of subject areas such as orthopedics, pediatrics, sports injuries, and nutrition, and are typically taught at chiropractic colleges or through professional associations. Finally, a growing number of institutions are offering accredited master's degrees in chiropractic-related fields. These degree programs are completed following either part-time (including hybrid and online learning) or full-time residential programs at institutions that offer the Doctor of Chiropractic degree program. Examples are Master of Science degrees in Nutrition/Functional Medicine and Sports and Exercise Science at the University of Western States, Applied Clinical Nutrition at New York Chiropractic College, Health Promotion at Cleveland Chiropractic College, Advanced Clinical Practice at National University of Health Sciences, Sports science and rehabilitation at Logan College of Chiropractic, and Master of Health Science degrees in Clinical Nutrition and Clinical Chiropractic Orthopedics at Northwestern Health Sciences University.

Accreditation

Each institution offering a Doctor of Chiropractic degree program in the United States is a member of the Association of Chiropractic Colleges (ACC) and is accredited by the Council on Chiropractic Education (CCE). CCE is the agency recognized by the United States Secretary of Education for accreditation of programs and institutions offering the Doctor of Chiropractic degree. CCE ensures the quality of chiropractic medical education in the US by means of accreditation, advancing educational improvement, and providing public information. CCE develops accreditation criteria to assess how effectively programs or institutions plan, implement and evaluate their mission and goals, program objectives, inputs, resources, and outcomes of their chiropractic programs.

The CCE is also recognized by the Council for Higher Education Accreditation (CHEA) and is a member of the Association of Specialized and Professional Accreditors (ASPA).[7] All but one of the institutions offering chiropractic programs in the United States also maintain institutional accreditation through regional higher education accrediting associations. Regional chiropractic accrediting bodies are also located in Canada (www.chirofed.ca), Europe (www.cce-europe.org), and Australasia (www.ccea.com.au).

Regulation and Certification

Regulatory status

Doctors of Chiropractic are regulated in all jurisdictions of the United States and Canada. Licensing boards from each state, and from many other jurisdictions outside the US, are members of the Federation of Chiropractic Licensing Boards (FCLB). The FCLB mission statement is "To protect the public and to serve our member boards by promoting excellence in chiropractic regulation."[8]

Licensing laws differ among the various jurisdictions. Some states such as Oregon, Illinois, Oklahoma, and New Mexico allow a broader scope of practice, while other states such as Washington and Michigan are more restrictive. Efforts are under way to develop and expand the scope of practice in most jurisdictions, to allow for the integration of contemporary practices and to better serve patients that seek care from chiropractic physicians. Members of jurisdictional licensing boards are generally governmental appointees and serve for established lengths of terms. Most boards also have public members that serve alongside the members of the chiropractic profession. More information on scope of practice can be found on the FCLB website. The FCLB also maintains a listing of actions taken against individual chiropractors in their Chiropractic Information Network-Board Action Databank (CIN-BAD), accessible on their website for a fee.

Similar organizations operate in Canada, Europe, and Australia. Local jurisdiction regulations also vary in each of these countries.

Chiropractic medicine is also practiced in many countries where official legal recognition has not yet occurred; in unregulated locations, those claiming to practice chiropractic often come from very diverse backgrounds and may lack any formal education or training beyond a few seminars. The World Federation of Chiropractic has been operational for over 20 years and has had significant influence and success in promoting legislation and standards for chiropractic practice.

Examinations and certifications

The National Board of Chiropractic Examiners (NBCE) is the principal testing agency for the chiropractic profession in the United States. Established in 1963, NBCE develops and administers standardized national examinations according to established guidelines. NBCE is dedicated to promoting excellence in the chiropractic profession by providing testing programs that measure educational attainment and clinical competency of those seeking licensure. Their examinations serve the needs of state licensing authorities, chiropractic colleges, educators and students, doctors of chiropractic, and the public. These examinations serve the profession and public by:

- promoting high standards of competence
- assisting state licensing agencies in assessing competence
- facilitating the licensure of newly graduated chiropractors, thereby enhancing professional credibility

In providing standardized written and performance assessments for the chiropractic profession, the NBCE develops, administers, analyzes, scores, and reports results from various examinations. The NBCE scores are among the criteria utilized by state licensing agencies to determine whether applicants demonstrate competency and satisfy state qualifications for licensure. NBCE scores are also utilized by the Council on Chiropractic Education accrediting agency as a recognized student learning outcome.

In its expanding role as an international testing agency, the NBCE espouses no particular chiropractic philosophy, but formulates test plans according to information provided collectively by the chiropractic colleges, the state licensing agencies, field practitioners, subject specialists, and a *Job Analysis of Chiropractic.*[9] There are four national exams students take in order to apply for licensure (most states accept the NBCE exam as their basic examination for licensure). The Part IV Exam is a practical clinical exam designed to test for competency and has gained international attention from other disciplines.[10]

Research

The Foundation for Chiropractic Education and Research (FCER) was formed in 1944 as the profession's foremost agency for funding of postgraduate scholarship and research. With the increased availability of federal grants to fund major research projects and the declining contributions of large sums of money from various sources, the FCER decided to cease operations in 2009. Chiropractic researchers at Palmer Chiropractic College, University of Western States, Northwestern Health Sciences University, Southern California University of Health Sciences, National University of Health Sciences, and New York Chiropractic College are among those in the United States who have received significant federal grants.

Palmer Chiropractic College has operated a National Institutes of Health (NIH) consortium center program for the last ten years, and has been involved with organizing the chiropractic profession's Research Agenda Conference (RAC). This meeting has been partially supported with federal grants. The Association of Chiropractic Colleges (ACC) sponsors annual conferences in which educators from around the world meet and present research papers and posters. Since the 1990s, this annual meeting has been combined with the RAC. These joint ACC-RAC conferences are attended by the academic and research community of chiropractic, as well as chiropractic practitioners, basic scientists and researchers from related medical and health care professions. Roughly $30 million in NIH grants was awarded to chiropractic colleges and universities from 1999-2016.

A significant portion of industry support for the profession has also been funneled into research, including over $10 million from the

National Chiropractic Mutual Insurance Company (NCMIC). Funds from Foot Levelers, Inc. have supported numerous fellowships, helping Doctors of Chiropractic obtain graduate research training at the Master's and PhD levels. Through such fellowships over the past 40 years, more than 100 Doctors of Chiropractic have earned graduate degrees (Master and Doctorate) in academic disciplines at major universities.

For the last ten years, the ACC has partnered with the World Federation of Chiropractic (WFC) to play an important role in sponsoring biennial international research symposia, both in the US and internationally. These events have provided venues for research and educational specialists on an international level to collaboratively present and discuss new developments in chiropractic health care and education. The WFC provides a forum for dialogue and exchange on a world-wide basis and has been productive in establishing common ground for the profession in individual countries around the world as well as with the World Health Organization.

At the federal level in the US, within the structure of the National Institutes of Health (NIH), the formation of what started as the Office for Alternative Medicine grew to become the National Center for Complementary and Alternative Medicine (NCCAM) and in 2015 was renamed the National Center for Complementary and Integrative Health (NCCIH). The change from having no presence at the NIH, to the formation of an Office with a budget of $2 million, to the formation of a Center with a budget in excess of $100 million is indeed evidence of dramatic growth. The chiropractic profession has played a significant role in the growth and recognition of complementary and integrative medicine through its research, educational, and legislative initiatives. A Doctor of Chiropractic was the first member of a complementary and integrative profession to be hired as program officer by the NIH.

Scientific journals

The ACC publishes the peer reviewed *Journal of Chiropractic Education*, which is indexed in PubMed. National University of Health Sciences (NUHS), through Elsevier Publishing, supports the publication of three peer-reviewed and indexed journals: *Journal of Manipulative and*

Physiological Therapeutics (JMPT), *Journal of Chiropractic Medicine* (JCM), and an online journal, *Journal of Chiropractic Humanities* (JCH). NUHS has published JMPT since 1978.

Challenges and Opportunities

Key challenges 2016 – 2020

- further expansion in federal healthcare delivery systems; healthcare reform; and agreements between chiropractic educational institutions, Department of Defense locations, and Department of Veterans Affairs facilities
- developing programs to improve enrollment trends at chiropractic educational institutions
- providing funding for research programs, residencies, and fellowships at all chiropractic institutions
- implementing measure to address the high cost of education
- exploring measures to provide fair compensation for Doctors of Chiropractic in clinical practice
- developing more opportunities for integration and collaboration with other healthcare professions
- better informing the public about the benefits of chiropractic health care
- adapting the practice of chiropractic to the needs of the public and considering realities of practicing evidence-informed care in a new healthcare reform environment, where cost, outcomes, and accountability are key elements

Key opportunities 2016-2020

- changes to the US healthcare delivery system with the inclusion and integration of chiropractic medical services on a broader scale
- providing the public with a better understanding of the education and training of the doctor of chiropractic and the safety and cost efficiencies conservative care contributes to healthcare

- continued growth and acceptance of complementary and integrative health care by the public and interprofessional education
- ongoing and increased research support through federal agencies such as NIH
- continued development of chiropractic education and integration within larger university systems
- more collaborative work among integrated healthcare professions
- continued growth and expansion of the chiropractic profession on a worldwide basis

Resources

Organizations and websites

In the US, there are two national organizations representing chiropractors: the American Chiropractic Association and the International Chiropractors Association. The World Federation of Chiropractic represents the profession on a global basis. There are approximately 100,000 chiropractic practitioners worldwide.

- American Chiropractic Association
  www.acatoday.org
- Association of Chiropractic Colleges
  www.chirocolleges.org
  www.DiscoverChiropractic.org
- Council on Chiropractic Education
  www.cce-usa.org
- Federation of Chiropractic Licensing Boards
  www.fclb.org
- International Chiropractors Association
  www.chiropractic.org
- National Board of Chiropractic Examiners
  www.nbce.org
- World Federation of Chiropractic
  www.wfc.org

Bibliography

Christensen MG, Kollasch MW. *Job Analysis of Chiropractic 2005.* Greeley, CO: National Board of Chiropractic Examiners; 2005.

Haldeman S, ed in chief. *Principles and Practice of Chiropractic.* 2nd ed. New York, NY: McGraw-Hill; 2004.

Leach RA. *The Chiropractic Theories: A Textbook of Scientific Research.* 4th ed. Baltimore, MD: Lippincott Williams & Wilkins; 2004.

Peterson DH, Bergmann TF. *Chiropractic Technique.* 2nd ed. St. Louis, MO: Mosby; 2000.

Phillips, RB. The chiropractic paradigm. *J Chiropr Educ.* 2001; 15(2):49-52.

Citations

1. Council on Chiropractic Education in the United States. Accreditation Standards – Principles, Processes & Requirements for Accreditation. January 2013. http://www.cce-usa.org/Publications.html
2. World Health Organization. *WHO Guidelines on basic training and safety in chiropractic.* World Health Organization. Geneva 2005. http://www.who.int/medicines/areas/traditional/Chiro-Guidelines.pdf
3. Key Facts About the Chiropractic Profession. American Chiropractic Associate website. https://www.acatoday.org/Patients/Why-Choose-Chiropractic/Key-Facts. Accessed February 16, 2017
4. The Current Status of the Chiropractic Profession. World Federation of Chiropractic. December 2012. https://www.wfc.org/website/images/wfc/WHO_Submission-Final_Jan2013.pdf.
5. www.salary.com
6. Chiropractic in the NFL. Professional Football Chiropractic Society website. http://profootballchiros.com/chiropractic-in-the-nfl/
7. Council on Chiropractic Education in the United States. www.cce-usa.org
8. Federation of Chiropractic Licensing Boards. www.fclb.org

9. National Board of Chiropractic Examiners. *The 2005 Job Analysis.* National Board of Chiropractic Examiners. Greeley, Colorado, 2005.
10. National Board of Chiropractic Examiners. www.nbce.org

Direct-Entry Midwifery

Third Edition (2017) Editors: Courtney L. Everson, PhD;
Nichole Reding, MA, CPM, LDM

First Edition (2009) Author and Second Edition (2013) Editor:
Jo Anne Myers-Ciecko, MPH

Third Edition (2017) Partner Organization:
Association of Midwifery Educators (AME)

First Edition (2009) and Second Edition (2013) Partner Organization:
Midwifery Education Accreditation Council (MEAC)

About the Authors/Editors: Myers-Ciecko is Senior Advisor for the Midwifery Education Accreditation Council. Everson is an ACIH Board Member, an Association of Midwifery Educators Board member, a Midwifery Education Accreditation Council Board member, the Director of Research Education for the Midwives Alliance of North America Division of Research, and a Medical Anthropologist and Dean of Graduate Studies at Midwives College of Utah. Reding is an ACIH Board Member, President of the Midwifery Education Accreditation Council, and Academic Coordinator at Birthingway College of Midwifery.

Philosophy, Mission, and Goals

Midwives are autonomous clinical practitioners that work in partnership with families to provide support, education, and care during the pregnancy, labor/birth, and postpartum periods. This care includes preventive measures, the promotion of normal physiologic birth, the detection of complications in mother and child, the accessing of medical or other appropriate assistance, and the carrying out of emergency measures. Midwives have an important task in health counseling and education, not only for individual clients, but also within the family and community. This work involves prenatal education and preparation for parenthood, and may extend to women's health, sexual or reproductive health, and childcare.[1] Direct-entry midwifery refers to an educational/training pathway that allows individuals to enter directly into the midwifery profession without first attending nursing school. Direct-entry Midwives may choose to obtain national credentialing as a Certified Professional Midwives (CPMs). The CPM credential is a national certification, first awarded in 1994 and issued through the North American Registry of Midwives. The National Association of Certified Professional Midwives (NACPM) was founded in 2000 to increase women's access to midwives by supporting the work and practice of CPMs. NACPM adopted the following *Philosophy and Principles of Practice:*[2]

1. NACPM Members respect the mystery, sanctity, and potential for growth inherent in the experience of pregnancy and birth.
2. NACPM members understand birth to be a pivotal life event for mother, baby, and family. It is the goal of midwifery care to support and empower the mother and to protect the natural process of birth.
3. NACPM members respect the biological integrity of the processes of pregnancy and birth as aspects of a woman's sexuality.
4. NACPM members recognize the inseparable and interdependent nature of the mother-baby pair.
5. NACPM members believe that responsible and ethical midwifery care respects the life of the baby by nurturing and respecting the mother, and, when necessary, counseling and educating her in ways to improve fetal/infant well-being.
6. NACPM members work as autonomous practitioners, recognizing that this autonomy makes possible a true partnership with the women they serve, and enables them to bring a broad range of skills to the partnership.
7. NACPM members recognize that decision making involves a synthesis of knowledge, skills, intuition and clinical judgment.
8. NACPM members know that the best research demonstrates that out-of-hospital birth is a safe and rational choice for healthy women, and that the out-of-hospital setting provides optimal opportunity for the empowerment of the mother and the support and protection of the normal process of birth.

9. NACPM members recognize that the mother or baby may on occasion require medical consultation or collaboration.
10. NACPM members recognize that optimal care of women and babies during pregnancy and birth takes place within a network of relationships with other care providers who can provide service outside the scope of midwifery practice when needed.

While NACPM represents CPMs, specifically, the Midwives Alliance of North America (MANA), founded in 1982, is a broad-based alliance representing the breadth and diversity of the profession of midwifery in the United States. Members include Certified Professional Midwives as well as Certified Nurse-Midwives (CNMs), state-licensed midwives, and traditional midwives who serve special populations such as the Amish or indigenous communities.

MANA "Core Competencies for Basic Midwifery Practice," last revised in 2014, are prefaced with the following guiding principles:[3]

> The midwife provides care according to the following guiding principles of practice:
>
> - Pregnancy and childbearing are natural physiologic life processes.
> - The biological wisdom to give birth is innate, it has been held throughout time, and is experienced across cultures by all pregnant people.
> - Physical, emotional, psychosocial and spiritual factors synergistically shape the health of individuals and affect the childbearing process.
> - The childbearing experience and birth of a baby are personal, family and community events.
> - Pregnant individuals are the only direct care providers for themselves and their unborn babies, thus the most important determinant of a healthy pregnancy is the pregnant person.
> - The parameters of "normal" vary widely, and each pregnancy, birth and baby is unique.

In consideration thereof:

- Midwives work in partnership with clients and their chosen support community throughout the caregiving relationship.
- Midwives respect and support the dignity, rights and responsibilities of the clients they serve.
- Midwives are committed to addressing inequities in health care status and outcomes.
- Midwives work as autonomous practitioners, and they collaborate with other health care and social service providers whenever appropriate.
- Midwives work to optimize the well-being of the mother-baby unit as the foundation of caregiving.
- Midwives recognize the empowerment inherent in the childbearing experience and strive to support clients to make informed decisions and take responsibility for their own and their baby's well-being.
- Midwives integrate clinical or hands-on evaluation, theoretical knowledge, intuitive assessment, spiritual awareness and informed consent and refusal as essential components of effective decision-making.
- Midwives strive to ensure optimal birth for the whole family and provide guidance, education and support to facilitate the spontaneous processes of pregnancy, labor and birth, lactation and mother–baby attachment, using appropriate intervention as needed.
- Midwives value continuity of care throughout the childbearing cycle and strive to maintain such continuity.
- Midwives are committed to sharing their knowledge and experience through such avenues as peer review, preceptorship, mentoring and participation in MANA's statistics collection program.

Characteristics and Data

There are approximately 2,900 Certified Professional Midwives (CPMs) (NARM, email communication, November 2016). Information on the number of direct-entry midwives who are not CPMs is not available. CPMs are typically self-employed and work in their community through home- and freestanding birth center settings. For more information on the practitioner and practice characteristics of CPMs, please see Cheyney et al. 2015.[4]

Clinical Care

Approach to patient care

Midwifery care encompasses the normal childbearing cycle of pregnancy, birth, and postpartum. NACPM, MANA, and other leading organizations that support direct-entry midwifery endorse the following statement:

> The *Midwives Model of Care*™ is based on the fact that pregnancy and birth are normal life events. The Midwives Model of Care includes:[5]

- monitoring the physical, psychological and social well-being of the mother throughout the childbearing cycle;
- providing the mother with individualized education, counseling, and prenatal care, continuous hands-on assistance during labor and delivery, and postpartum support;
- minimizing technological interventions; and
- identifying and referring women who require obstetrical attention.

Research demonstrates that midwifery models of care can decrease medical interventions, such as cesareans, while achieving positive birth outcomes for both the mother and baby.[5-15]

Researchers who interviewed direct-entry midwives early in the development of the home-birth movement found that a wellness orientation, shared responsibility, passive management, holistic care, and individualized care were central to the midwifery approach to care.[16] Exemplary midwifery practice was described by clients, direct-entry midwives, and nurse-midwives in another study that identified critical process-of-care qualities, such as supporting the normalcy of birth, respecting the uniqueness of the woman and family, vigilance and attention to detail, and creating a setting that is respectful and reflects the family's needs.[17]

The North American Registry of Midwives (NARM) states that, "CPMs work with women to promote a healthy pregnancy, and provide education to help her make informed decisions about her own care. In partnership with their clients they carefully monitor the progress of the pregnancy, labor, birth, and postpartum period and recommend appropriate management if complications arise, collaborating with other healthcare providers when necessary. The key elements of this education, monitoring, and decision making process are based on evidence-based practice and informed consent."[18]

Scope of practice

NACPM defines the midwives' scope of practice as providing expert care, education, counseling, and support to women and their families throughout the caregiving partnership, including pregnancy, birth, and postpartum. NACPM members provide ongoing care throughout pregnancy and continuous, hands-on care during labor, birth, and the immediate postpartum period. They are trained to recognize abnormal or dangerous conditions needing expert help outside their scope and to consult or refer as necessary.

NARM recognizes that each midwife is an individual with specific practice protocols that reflect her own style and philosophy, level of experience, and legal status, and that practice guidelines may vary with each midwife. NARM does not set protocols for all CPMs to follow, but requires that they develop their own practice guidelines in written form as part of certification.

In certain jurisdictions, the midwives' scope of practice includes well-woman care and family planning services. Midwives may also administer certain drugs and devices as specified in state law, including IV fluids, antibiotics, local anesthetic, antihemorrhagics for postpartum use, etc. Midwives typically carry oxygen and resuscitation equipment and are certified in adult and neonatal resuscitation.

Referral practices

Direct-entry midwives care for essentially healthy clients expecting normal pregnancy, birth, and postpartum experiences. Midwives consult with, collaborate with, and refer to an array of healthcare professionals, social service providers, and others whose services may benefit the families they serve. These include, but are not limited to, allopathic physicians, naturopathic physicians, acupuncturists, chiropractors, massage therapists, childbirth educators, psychologists, family therapists, doulas, nutritionists, social workers, and food support and housing agencies. In some states, midwives are legally obligated to consult with and/or refer women to obstetricians or family physicians with obstetrical privileges when certain conditions arise. In addition, because the vast majority of direct-entry midwives attend births in their clients' homes or in freestanding birth centers, any complications that require hospitalization for labor and/or birth necessitate transfer of care to an obstetrical care provider with hospital privileges. Most transfers are non-emergent, resulting from a failure to progress in labor, the mother's desire for pain relief, or maternal exhaustion. Less common, but more urgent indications for transfer include preeclampsia, maternal hemorrhage, retained placenta, malpresentation, sustained fetal distress, and respiratory problems in the newborn.

Third-party payers

Direct-entry midwives are reimbursed by private insurance plans and contract with managed care organizations in many states. A number of states mandate reimbursement under "every category of provider" or "any willing provider" laws. At least ten states reimburse direct-entry midwives for services provided to women on Medicaid. This is a high-priority issue for NACPM, which is committed to achieving national recognition for the CPM profession, including mandatory inclusion in Medicaid programs. The Patient Protection and Affordable Care Act of 2010 stipulates Medicaid reimbursement for licensed care providers working in birth centers, which includes licensed midwives.

Integration Activities

NACPM is represented in numerous national organizations and initiatives focused on healthcare reform and women's health issues. These include the Integrated Healthcare Policy Consortium, Health Care for America Now, and the National Quality Forum. MANA is a member of the Coalition for Improving Maternity Services, a broad-based coalition of over 50 organizations, representing over 90,000 members. The coalition's mission is to promote a wellness model of maternity care that will improve birth outcomes and substantially reduce costs. MANA is also a partner in The Safe Motherhood Initiatives-USA, a partnership of organizations whose goal is to reduce maternal mortality in the United States. NACPM and MANA joined with four other groups to form the Midwives and Mothers in Action Campaign, a partnership whose goal is to gain federal recognition of certified professional midwives so that women and families will have increased access to quality, affordable maternity care in the settings of their choice.

In 2012, NACPM and the Association of Midwifery Educators co-hosted a symposium focused on the opportunities for integration and expansion of midwifery in a new maternity care system. A series of reports from the symposium were issued throughout 2012 and are available on the CPM Symposium website.

At the Home Birth Summit in 2011, a diverse group of stakeholders, including midwives, physicians, researchers and policy-makers, were invited to examine home birth within the context of the maternity care system. Using Future Search methodology to identify common ground and build consensus, the goal was to establish what the whole system can do to support those who choose homebirth, and provide the care, safety net, consultation, collaboration and referral necessary to make homebirth the safest and most positive experience for all involved: moms, babies, families, communities, healthcare workers, hospital personnel, administrators, payers, and so on.[19]

The following statements reflect the areas of consensus that were achieved by the individuals who participated in the Home Birth Summit: [20]

STATEMENT 1

We uphold the autonomy of all childbearing women. All childbearing women, in all maternity care settings, should receive respectful, woman-centered care. This care should include opportunities for a shared decision-making process to help each woman make the choices that are right for her. Shared decision making includes mutual sharing of information about benefits and harms of the range of care options, respect for the woman's autonomy to make decisions in accordance with her values and preferences, and freedom from coercion or punishment for her choices.

STATEMENT 2

We believe that collaboration within an integrated maternity care system is essential for optimal mother-baby outcomes. All women and families planning a home or birth center birth have a right to respectful, safe, and seamless consultation, referral, transport and transfer of care when necessary. When ongoing interprofessional dialogue and cooperation occur, everyone benefits.

STATEMENT 3

We are committed to an equitable maternity care system without disparities in access, delivery of care, or outcomes. This system provides culturally appropriate and affordable care in all settings, in a manner that is acceptable to all communities. We are committed to an equitable educational system without disparities in access to affordable, culturally appropriate, and acceptable maternity care provider education for all communities.

STATEMENT 4
It is our goal that all health professionals who provide maternity care in home and birth center settings have a license that is based on national certification that includes defined competencies and standards for education and practice. We believe that guidelines should:

- allow for independent practice
- facilitate communication between providers and across care settings
- encourage professional responsibility and accountability; and,
- include mechanisms for risk assessment.

STATEMENT 5
We believe that increased participation by consumers in multi-stakeholder initiatives is essential to improving maternity care, including the development of high quality home birth services within an integrated maternity care system.

STATEMENT 6
Effective communication and collaboration across all disciplines caring for mothers and babies are essential for optimal outcomes across all settings. To achieve this, we believe that all health professional students and practitioners who are involved in maternity and newborn care must learn about each other's disciplines, and about maternity and health care in all settings.

STATEMENT 7
We are committed to improving the current medical liability system, which fails to justly serve society, families, and health-care providers and contributes to:

- inadequate resources to support birth injured children and mothers

- unsustainable healthcare and litigation costs paid by all
- a hostile healthcare work environment
- inadequate access to home birth and birth center birth within an integrated healthcare system; and
- restricted choices in pregnancy and birth.

STATEMENT 8
We envision a compulsory process for the collection of patient (individual) level data on key process and outcome measures in all birth settings. These data would be linked to other data systems, used to inform quality improvement, and would thus enhance the evidence basis for care.

STATEMENT 9
We recognize and affirm the value of physiologic birth for women, babies, families and society and the value of appropriate interventions based on the best available evidence to achieve optimal outcomes for mothers and babies.

In 2006, the White Ribbon Alliance for Safe Motherhood convened a national working group to develop guidelines to ensure that the healthcare needs of pregnant women, new mothers, fragile newborns, and infants would be adequately met during and after a disaster. They recommended that CPMs should be engaged in local and regional planning efforts, that home-birth skills should be taught to all providers, and that information be provided on how to prepare for birth outside the hospital.

In 2001 the American Public Health Association (APHA) adopted a resolution supporting "Increasing Access to Out-Of-Hospital Maternity Care Services through State-Regulated and Nationally-Certified Direct-Entry Midwives." APHA supports licensing and certification for direct-entry midwives, increased funding for scholarship and loan repayment programs, and eliminating barriers to the reimbursement and equitable payment of direct-entry midwives.[21]

In 1999 the Pew Health Professions Commission and the UCSF Center for the Health Professions issued a joint report on "The Future of Midwifery" calling for expanding educational opportunities for nurse-midwifery and direct-entry midwifery, health policies that facilitate integration of midwifery services, and research to evaluate practices.[22]

Education

Schools and Programs

There are currently ten schools and programs accredited by the Midwifery Education Accreditation Council (MEAC) with more aspiring schools seeking accreditation regularly. Of the 10 current schools, three offer Master's degrees, four offer associate or bachelor's degrees, and six offer certificates in midwifery. All schools combine didactic academic coursework with clinical preceptorships. Several programs are US Department of Education Title IV approved schools.

There are nearly 600 students enrolled in accredited midwifery programs and institutions. In addition, there are at least that many more independent students who are completing community-based apprenticeships, following guidelines provided by NARM, whose competency will be assessed individually through NARM's Portfolio Evaluation Process.

Accredited programs (October 2016):

- Bastyr University, Department of Midwifery, Kenmore, WA http://www.bastyr.edu/academics/schools-departments/school-natural-health-arts-sciences/department-midwifery
- Southwest Wisconsin Technical College Midwifery Program, Fennimore, WI https://www.swtc.edu/academics/programs/health-occupations/midwife

Accredited institutions (October 2016):

- Birthingway College of Midwifery, Portland, OR http://www.birthingway.edu/

- Birthwise Midwifery School, Bridgton, ME
  http://www.birthwisemidwifery.edu/
- Florida School of Traditional Midwifery, Gainesville, FL
  http://www.midwiferyschool.org
- Maternidad La Luz, El Paso, TX
  http://www.maternidadlaluz.com/
- Midwives College of Utah, Salt Lake City, UT
  http://www.midwifery.edu/
- National College of Midwifery, Taos, NM
  http://www.midwiferycollege.org/
- National Midwifery Institute, Bristol, VT
  http://www.nationalmidwiferyinstitute.com/
- Nizhoni Institute of Midwifery, San Diego, CA
  http://www.midwiferyatnizhoni.com/

Curriculum content

The MANA Core Competencies establish the essential knowledge, clinical skills and critical thinking necessary for entry-level practice for direct-entry midwifery in the United States. The Certified Professional Midwife (CPM) credential is based on the MANA Core Competencies.

The MANA Core Competencies were written and adopted by the MANA Board of Directors on October 3, 1994, and revised and adopted on August 4, 2011 and again in December 2014. The MANA Core Competencies are:[23]

I. General Knowledge and Skills

The midwife's knowledge and skills include but are not limited to:

A. communication, counseling and education;
B. human anatomy and physiology;
C. human sexuality;
D. various therapeutic health care modalities for treating body, mind and spirit;
E. community health care, wellness and social service resources;
F. nutritional needs, health and lifestyle habits;
G. diversity awareness, sensitivity and competency.

The midwife's knowledge and skills relate community health to client needs, including but not limited to:

A. the community and social determinants of health, including race, income, literacy and education, water supply and sanitation, housing, environmental hazards, food security, disease patterns, and common threats to health;
B. principles of community-based primary care using health promotion and disease prevention and control strategies;
C. direct and indirect causes of maternal and neonatal mortality and morbidity in the local community, and strategies for reducing them;
D. principles of epidemiology;
E. principles of health education;
F. emergency preparedness for disaster response including communication and transport mechanisms;
G. human rights and their effects on health of individuals, including issues such as domestic violence, genital circumcision, gender equity, gender identity and expression, and how their expression affects health outcomes;
H. advocacy and empowerment strategies;
I. culture and beliefs, including religion, social norms, family structure and health practices;
J. birth planning, benefits and risks of available birth settings.

The midwife maintains professional standards of practice including but not limited to:

A. principles of informed consent and refusal and shared decision making;
B. critical evaluation of evidence-based research findings and application to best practices;
C. documentation of care throughout the childbearing cycle;
D. consistent actions in accordance with professional ethics, values and human rights;

E. courteous, non-judgmental, non-discriminatory, and culturally appropriate behaviors with all clients;
F. respect for individuals and their culture, customs and beliefs, ethnic origin, gender identity, sexual orientation, family structure, and religious beliefs;
G. knowledge of commonly used medical terminology;
H. implementation of individualized plans for client-centered midwifery care;
I. support for the relationship among the mother-baby unit, the family and their larger community;
J. judicious use of technology;
K. self-assessment and acknowledgement of personal and professional limitations.

II. Care during Pregnancy

The midwife provides care, support and information throughout pregnancy and determines the need for consultation, referral or transfer of care as appropriate. The midwife has knowledge and skills to provide care that include but are not limited to:

A. identification, evaluation and support for the client's and baby's well-being throughout the process of pregnancy;
B. initial and ongoing history at each antenatal visit;
C. physical examination and explanation of findings to the client;
D. education and counseling during the childbearing cycle;
E. identification of pre-existing conditions and preventive or supportive measures to enhance client well-being during pregnancy;
F. nutritional requirements of pregnancy and methods of nutritional assessment and counseling;
G. emotional, psychosocial and sexual variations that may occur during pregnancy;
H. environmental and occupational hazards during pregnancy;

I. effects of smoking, alcohol and drug use on pregnancies and unborn babies;
J. methods of diagnosing pregnancy;
K. the growth and development of the unborn baby;
L. genetic factors that may indicate the need for counseling, testing, or referral;
M. screening methods and diagnostic tests used during pregnancy, including indications, risks and benefits;
N. health and psychosocial needs associated with spontaneous or therapeutic abortion, including referral to community resources;
O. anatomy, physiology, and evaluation of the soft and bony structures of the pelvis;
P. palpation skills for evaluation of the baby and the uterus;
Q. the causes, assessment and treatment of the common discomforts of pregnancy;
R. Identification, implications and appropriate treatment of various infections, disease conditions and other problems that may affect pregnancy;
S. basic principles of pharmacokinetics of drugs prescribed, dispensed or administered during pregnancy;
T. effects of prescribed medications, herbal medicines, and over-the-counter drugs on pregnancy and the baby;
U. administration of medications as indicated;
V. management and care of the Rh-negative client;
W. signs, symptoms and indications for referral of selected complications and conditions of pregnancy;
X. the physiology of lactation and methods to prepare for breastfeeding;
Y. counseling to the client and family to plan for a safe, appropriate place of birth.

III. Care During Labor, Birth and the Immediate Postpartum

The midwife provides care, support and information throughout labor, birth and the hours immediately thereafter. The midwife determines the need for consultation, referral or transfer of care as appropriate. The midwife has knowledge and skills to provide care that include but are not limited to:

A. the processes of labor and birth;
B. parameters and methods, including relevant health history, for evaluating the client's and baby's well-being during labor, birth and immediately thereafter;
C. assessment of the birthing environment to assure that it is clean, safe and supportive and that appropriate equipment and supplies are on hand;
D. attention to emotional responses and their impact during labor, birth and immediately thereafter;
E. comfort and support measures during labor, birth and immediately thereafter;
F. fetal and maternal anatomy and their interrelationship as relevant to assessing the baby's position and the progress of labor;
G. hydration and nutritional requirements during labor, birth and immediately thereafter;
H. techniques to assist and support the spontaneous vaginal birth of the baby and placenta;
I. recommendations for rest and sleep as appropriate during the process of labor, birth and immediately thereafter;
J. techniques to assist and support labor, birth and the immediate postpartum in water;
K. treatment for variations that can occur during the course of labor, birth, and immediately thereafter, including prevention and treatment of maternal hemorrhage;
L. emergency measures and transport for critical problems arising during labor, birth, or immediately thereafter;
M. appropriate support for the newborn's natural physiologic transition during the first minutes and hours following birth, including skin-to-skin contact and practices to enhance mother-baby attachment and family bonding;

N. pharmacological measures for management and control of indications in the intrapartum and immediate postpartum for client and baby;
O. current interventions and technologies that may be commonly used in a medical setting;
P. care and repair of the perineum and surrounding tissues;
Q. third-stage management, including assessment of the placenta, membranes and umbilical cord, and collection of the cord blood;
R. breastfeeding and lactation;
S. identification of pre-existing conditions and implementation of preventive or supportive measures to enhance client well-being during labor, birth, the immediate postpartum and breastfeeding.

IV. Postpartum Care

The midwife provides care, support and information throughout the postpartum period and determines the need for consultation, referral or transfer of care as appropriate. The midwife has knowledge and skills to provide care that include but are not limited to:

A. anatomy and physiology of the birthing parent;
B. lactation support and appropriate breast care and treatments for breastfeeding problems or complications, including mastitis;
C. management and care of the Rh-negative client with the Rh-positive baby;
D. support for the client's well-being and mother-baby attachment;
E. treatment for client discomforts;
F. nutrition, rest, activity and physiological needs during the postpartum period and lactation;
G. emotional, psychosocial, mental, and sexual variations;
H. signs and symptoms of postpartum conditions requiring management, including those needing immediate medical intervention;

I. current identification and treatments for psychosocial adjustment problems including postpartum depression and mental illness;
J. principles of interpersonal communication with, and support for, grief counseling when necessary;
K. family planning methods, as desired.

V. Newborn Care
The midwife provides care to the newborn during the postpartum period, as well as support and information to parents regarding newborn care and informed decision making, and determines the need for consultation, referral or transfer of care as appropriate. The midwife's assessment, care and shared information include but are not limited to:

A. anatomy, physiology and support of the newborn's adjustment during the first days and weeks of life;
B. newborn wellness, including relevant historical data and gestational age;
C. nutritional needs of the newborn;
D. benefits of breastfeeding and lactation support;
E. prophylactic treatments and screening tests commonly used during the neonatal period including applicable laws and regulations;
F. newborn growth, development, behavior, nutrition, feeding, and care;
G. traditional or cultural practices related to the newborn;
H. neonatal problems and abnormalities, and referral as appropriate;
I. discussion of circumcision and immunizations;
J. safety needs of the newborn.

VI. Women's Health Care and Family Planning
The midwife provides care, support and information regarding reproductive health and determines the need for consultation or

referral by using a foundation of knowledge and skills that includes but is not limited to:

A. reproductive health care across the lifespan;
B. evaluation of the client's well-being, including relevant health history;
C. common laboratory tests and screenings;
D. physical examination, including clinical breast and pelvic examination, focused on the presenting condition of the client;
E. anatomy and physiology related to conception and reproduction;
F. contemporary family planning methods, including natural, chemical and surgical methods of contraception, mode of action, indications, benefits and risks;
G. decision making regarding timing of pregnancies and resources for counseling and referral;
H. preconception and interconceptual care;
I. wellness care and gynecology.

VII. Professional, Legal and Other Aspects of Midwifery Care
The midwife assumes responsibility for practicing in accordance with the principles and competencies outlined in this document. The midwife uses a foundation of theoretical knowledge, clinical assessment, critical-thinking skills and shared decision making that are based on:

A. MANA's Essential Documents concerning the art and practice of midwifery,
B. the purpose and goals of MANA and local (state or provincial) midwifery associations,
C. principles and practice of data collection as relevant to midwifery practice,
D. ongoing education,
E. peer review, quality assessment and other professional and legal accountability processes;

F. principles of research, evidenced-based practice, critical interpretation of professional literature, and research findings;
G. professional guidelines, jurisdictional laws and regulations governing the practice of midwifery, health and reproduction;
H. knowledge of community health care delivery systems, and needed resources for midwifery care;
I. strategies for increasing access to midwifery care, especially in underserved communities;
J. skills in entrepreneurship and midwifery business management.

Accreditation

The Midwifery Education Accreditation Council (MEAC) is recognized by the US Secretary of Education as the accrediting agency for direct-entry midwifery education programs. All MEAC-accredited institutions and programs offer comprehensive education that prepare graduates to become CPMs. Most of the accredited programs average three years in length.

As stated by MEAC. "The Midwifery Education Accreditation Council's mission is to promote excellence in midwifery education through accreditation. It creates standards and criteria for the education of midwives. MEAC standards incorporate the nationally recognized core competencies and guiding principles set by the MANA, the International Confederation of Midwives (ICM), and the requirements for national certification of NARM. MEAC's accreditation criteria for midwifery education programs reflect the unique components and philosophy of the Midwives Model of Care."[24]

Regulation and Certification

Regulatory status

Direct-entry midwives are authorized to legally practice in 28 states with new legislation being proposed daily. Rules and regulations for practice vary state-by-state. Current information, including access to individual state laws, is available at the MANA website. [25]

In 2012, NARM adopted a position statement in support of state licensure for CPMs.[26] At that time, there were active efforts in at least 11 states to secure licensure legislation.

Examinations and certifications

NARM is the certifying agency for CPMs. Certification is a credential that validates the knowledge, skills, and abilities vital to responsible midwifery practice, and that reflects and preserves the essential nature of midwifery care. The CPM credential is unique among maternity care providers in the US as it requires training and experience in out-of-hospital birth. The NARM certification process recognizes multiple routes of entry into midwifery and includes verification of knowledge and skills as well as the successful completion of both a written examination and skills assessment. NARM conducts periodic surveys of CPMs and completes a national job analysis to assure that the examination is based on real-life job requirements. Candidates for certification must be graduates of an accredited midwifery program or must complete a Portfolio Evaluation Process administered by NARM. Certification is renewed every three years and all CPMs must obtain continuing education and participate in peer review for recertification.

The CPM credential is accredited by the National Commission for Certifying Agencies (NCCA), which is the accrediting body of the National Organization for Competency Assurance (NOCA). The Certified Nurse-Midwife credential is also accredited by the NCCA.

NCCA encourages their accredited certification programs to have an education evaluation process so candidates who have been educated outside of established pathways may have their qualifications evaluated for credentialing. The NARM Portfolio Evaluation Process meets this recommendation.

Certified professional midwives must demonstrate that they have met the minimum education, skills, and experience requirements set forth by NARM. Competence is assessed against a comprehensive and detailed list of knowledge, skills and abilities. Students must also participate in specific minimum numbers of prenatal, postpartum and newborn care experiences.

Research

There is a growing body of evidence that demonstrates the many benefits of midwife-led care in home- and birth center settings.[5-8,10-14,27-29] Most notably, in 2014, the largest study to-date in the United States on outcomes for midwife-led care in the homebirth setting was published, demonstrating positive perinatal health outcomes, including high rates of normal physiologic birth and very low rates of operative birth and interventions.[8] This published research was based on data from the MANA Statistics project, the largest perinatal data registry in the United States on midwife-led care in home and birth center settings. This ongoing, validated data registry[30] contributes not only to research on outcomes of care, but also on specific care practices in maternity care, such as vaginal birth after cesarean[31] and waterbirth.[32]

Notably, in 2016, a Cochrane Review on midwife-led continuity models of care concluded that "This review suggests that women who received midwife-led continuity models of care were less likely to experience intervention and more likely to be satisfied with their care with at least comparable adverse outcomes for women or their infants than women who received other models of care…Most women should be offered 'midwife-led continuity of care'. It provides benefits for women and babies and we have identified no adverse effects".[33]

Midwife models of care have also been implicated in improving perinatal health inequities, especially for families of color and reducing preterm delivery rates[34], as well as creating cost savings for the healthcare system and increasing access for low-income families.[10,27] Finally, a growing body of literature examines interprofessional collaboration and characteristics of Certified Professional Midwives.[35]

Challenges and Opportunities

Key challenges 2016—2020

- coordinate an effective response to the increasing medicalization of pregnancy and birth, including the rapidly rising cesarean section rate
- eliminate disparities in perinatal outcomes by addressing the role that midwives can play and promoting the benefits of midwifery care
- preserve and promote the principles, values, and practices of midwifery within the context of continuing professionalization of the home-birth and midwifery movements
- secure legislation in states where midwifery is not recognized or is currently illegal
- support faculty development, particularly pedagogical training for clinical and academic faculty
- increase the number of students of color and midwife instructors of color as a key strategy in increasing the number of midwives of color and their impact on equitable midwifery care
- build capacity and develop funding to support new and ongoing professional organization activities
- develop structures and standards to support competency-based and direct assessment programs and schools

Key opportunities 2016—2020

- public awareness and demand for midwifery services is increasing; support continued public education and outreach, including specific campaigns launched by MANA and the American College of Nurse-Midwives (ACNM)
- healthcare reform includes a number of federal initiatives to improve maternity care; leverage these to increase recognition, expand job opportunities and third party reimbursement for direct-entry midwives
- home birth services and birth centers are gaining recognition as a legitimate, high-quality and cost-effective element of maternity

care; build on the consensus statements from the 2011 Home Birth Consensus Summit to improve integration and support for women and their care providers in all settings
- workforce needs are changing and a shortage of maternity care providers is predicted; develop funding sources and other support to increase the number of direct-entry midwifery education programs and expand opportunities for training
- health professional education, like higher education more generally, is changing rapidly to meet the needs of a new era; encourage innovation to improve access to cutting-edge learning opportunities
- midwifery research opportunities are growing; support evaluation of midwifery practices to expand the knowledge base and to inform women's choices in maternity care and decisions by health policy-makers
- global standards for midwifery education and regulation have been set by the International Confederation of Midwives; utilize these tools to create common understanding and build unity among all midwifery organizations in the US

Resources

Organizations and websites

The Allied Midwifery Organizations (AMOs) are the leading national midwifery organizations that collaborate together to advance direct-entry midwifery[36], listed here in alphabetical order:
- Association of Midwifery Educators (AME), a non-profit organization committed to strengthening midwifery educators, schools, and administrators through connection, collaboration and coordination. www.associationofmidwiferyeducators.org/
- Citizens for Midwifery, a consumer-based grassroots organization promoting the Midwives Model of Care through education and outreach. www.cfmidwifery.org/
- International Center for Traditional Childbearing (ICTC), works to increase the number of Black midwives, doulas, and healers to

empower families, in order to reduce infant and maternal mortality. http://ictcmidwives.org/
- Midwifery Education Accreditation Council (MEAC), an independent, nonprofit organization recognized by the US Department of Education. http://meacschools.org/
- Midwives Alliance of North America (MANA), brings all midwives together regardless of route of entry into the profession, training, educational background or practice style. http://mana.org/
- National Association of Certified Professional Midwives (NACPM), a professional organization of CPMs formed to support the education, practice, and advancement of CPMs and inspire and engage them to be an organized force for change to increase access to high quality, high value maternity care for all women. http://nacpm.org/
- North American Registry of Midwives (NARM), sets standards for competency-based certification for Certified Professional Midwives (CPMs). http://narm.org/

Other midwifery organizations include:
- American Association of Birth Centers www.birthcenters.org/
- American College of Nurse-Midwives www.acnm.org/
- Childbirth Connection www.childbirthconnection.org/
- Coalition for Improving Maternity Services www.motherfriendly.org/
- CPM Symposium www.cpmsymposium.com/
- Foundation for the Advancement of Midwifery www.foundationformidwifery.org/
- International Confederation of Midwives www.internationalmidwives.org/

Bibliography

Selected books about midwifery and maternity care

Bridgman Perkins, B. *The Medical Delivery Business: Health Reform, Childbirth and the Economic Order.* Piscataway, NJ: Rutgers University Press; 2004.

Davis-Floyd R, Johnson CB, eds. *Mainstreaming Midwives: The Politics of Change.* New York, NY: Routledge; 2006.

Gaskin IM. *Ina May's Guide to Natural Childbirth.* New York, NY: Bantam; 2003.

Rooks JP. *Midwifery and Childbirth in America.* Philadelphia, PA: Temple University Press; 1999.

Rothman BK. *Laboring On: Birth in Transition in the United States.* New York, NY: Routledge; 2007.

Wagner M. *Born in the USA: How a Broken Maternity System Must Be Fixed to Put Women and Children First.* Berkeley, CA: University of California Press; 2006.

Clinical Reference texts

Cunningham F, Gant NF, Leveno KJ, Gilstrap LC, Hauth JC, Wenstrom, K. *William's Obstetrics.* 21st ed. New York, NY: McGraw-Hill; 2001.

Davis E. *Heart and Hands: A Midwife's Guide to Pregnancy and Birth.* 4th ed. Berkeley, CA: Celestial Arts; 2004.

Enkin M, Keirse M, Neilson J, et al. *A Guide to Effective Care in Pregnancy and Birth.* New York, NY: Oxford University Press; 2000.

Frye A. *Holistic Midwifery: A Comprehensive Textbook for Midwives and Home Birth Practice. Volume 1 – Care During Pregnancy.* Portland, OR: Labrys Press; 1995.

Frye A. *Holistic Midwifery: A Comprehensive Textbook for Midwives and Home Birth Practice. Volume 2 – Care During Labor and Birth.* Portland, OR: Labrys Press; 2004.

Frye A. *Understanding Diagnostic Tests in the Childbearing Year.* Portland, OR: Labrys Press; 1997.

Gaskin I. *Spiritual Midwifery.* 4th ed. Summertown, TN: The Book Publishing Company; 2002.

Myles M. *Textbook for Midwives.* 14th ed. Philadelphia, PA: Elsevier; 2000.
Oxhorn H. *Human Labor and Birth.* 5th ed. New York, NY: McGraw-Hill; 1986.
Page L. *The New Midwifery.* Philadelphia, PA: Churchill Livingstone; 2000.
Renfrew M, Fisher C, Arms S. *Bestfeeding: Getting Breastfeeding Right.* 2nd ed. Berkeley, CA: Celestial Arts; 2000.
Simkin P. *Labor Progress Handbook.* Cambridge, MA: Blackwell Scientific; 2000.
Sinclair C. *A Midwife's Handbook.* St. Louis, MO: Saunders; 2004.
Thureen P. *Assessment and Care of the Well Newborn.* St. Louis, MO: Saunders; 1998.
Varney H, Kriebs JM, Gegor CL. *Varney's Midwifery.* 4th ed. Sudbury, MA: Jones and Bartlett; 2004.
Weaver P, Evans S. *Practical Skills Guide for Midwifery.* 3rd ed. Wasilla, AK: Morningstar Publishing; 2001.
Wickham S. *Midwifery Best Practice.* Philadelphia, PA: Elsevier; 2003.

Reports and recommendations

Lamaze International. The Coalition for Improving Maternity Services: Evidence Basis for the Ten Steps of Mother-Friendly Care. *Supp J Perinat Educ.*2007;16(1).
The Mother-Friendly Childbirth Initiative. Coalition for Improving Maternity Services Web site. http://www.motherfriendly.org/mfci.php. 1996. Accessed November 17, 2009.
Wells S, Nelson C, Kotch JB, et al. Increasing Access to Out-of-Hospital Maternity Care Services through State-Regulated, Nationally-Certified Direct-Entry Midwives. Midwives Alliance of North America Website. http://mana.org/APHAformatted.pdf. 2001. Accessed November 17, 2009

Midwifery education, professional associations, state regulation

Bourgeault I, Fynes M. Integrating lay and nurse-midwifery into the US and Canadian healthcare systems. *Soc Sci Med.* 1997;44(7):1051-1063.

Suarez SH. Midwifery is not the practice of medicine. *Yale J Law Fem.* 1993;5(2):315-364.

Scope of practice, standards of practice, and quality assurance

Spindel PG, Suarez SH. Informed consent and home birth. *J Nurse Midwifery*. 1995;40(6):541-552.

Suarez SH. Midwifery is not the practice of medicine. *Yale J Law Fem.* 1993;5(2):315-364.

Consultation and referral

Davis-Floyd R. Home-birth emergencies in the US and Mexico: the trouble with transport. *Soc Sci Med*. 2003;56:1911-1931.

Stapleton SR. Team-building: making collaborative practice work. *J Nurse Midwifery*. 1998;43(1):12-18.

Citations

1. International Confederation of Midwives. International definition of the midwife. http://www.internationalmidwives.org/who-we-are/policy-and-practice/icm-international-definition-of-the-midwife/
2. NACPM Core Documents. 2014. NACPM website. http://nacpm.org/about-cpms/professional-standards/nacpm-core-documents/
3. Core Competencies for Basic Midwifery Practice. December 2014. Midwives Alliance North America. http://mana.org/about-us/core-competencies
4. Cheyney M, Olsen C, Bovbjerg M, et al. 2015. "Practitioner and Practice Characteristics of Certified Professional Midwives in the United States: Results of the 2011 North American Registry of Midwives Survey." *Journal of Midwifery & Women's Health* 60 (5): 534–45. doi:10.1111/jmwh.12367.
5. The Midwives Model of Care. Citizens for Midwifery website. http://cfmidwifery.org/mmoc/define.aspx
6. Bolten N, de Jonge E, Zwagerman P, et al. Effect of Planned Place of Birth on Obstetric Interventions and Maternal Outcomes among Low-Risk Women: A Cohort Study in the Netherlands.

BMC Pregnancy and Childbirth. 2016;16 (1): 329. doi:10.1186/s12884-016-1130-6.

7. Brocklehurst P, Hardy P, Hollowell, J, et al. Perinatal and Maternal Outcomes by Planned Place of Birth for Healthy Women with Low Risk Pregnancies: The Birthplace in England National Prospective Cohort Study." *BMJ (Clinical Research Ed.).* 2011;343: d7400.
8. Cheyney M, Bovbjerg M, Everson C, et al. Outcomes of Care for 16,924 Planned Home Births in the United States: The Midwives Alliance of North America Statistics Project, 2004 to 2009. *Journal of Midwifery & Women's Health* 2014;59 (1): 17–27. doi:10.1111/jmwh.12172.
9. Fullerton J, Navarro A, and Young S. Outcomes of Planned Home Birth: An Integrative Review. *Journal of Midwifery & Women's Health.* 2007;52 (4): 323–33. doi:10.1016/j.jmwh.2007.02.016.
10. Hendrix M, Evers S, Basten M, et al. Cost Analysis of the Dutch Obstetric System: Low-Risk Nulliparous Women Preferring Home or Short-Stay Hospital Birth--a Prospective Non-Randomised Controlled Study. *BMC Health Services Research.* 2009;9: 211. doi:10.1186/1472-6963-9-211.
11. Hutton E, Reitsma A, Kaufman K. Outcomes Associated with Planned Home and Planned Hospital Births in Low-Risk Women Attended by Midwives in Ontario, Canada, 2003-2006: A Retrospective Cohort Study. *Birth.* 2009;36 (3): 180–89. doi:10.1111/j.1523-536X.2009.00322.x.
12. Janssen P, Lee S, Ryan E, Etches D, et al. Outcomes of Planned Home Births versus Planned Hospital Births after Regulation of Midwifery in British Columbia. *CMAJ.* 2002;166 (3): 315–23.
13. Janssen P, Saxell L, Page L, et al. Outcomes of Planned Home Birth with Registered Midwife versus Planned Hospital Birth with Midwife or Physician. *CMAJ.* 2009;181 (6–7): 377–83. doi:10.1503/cmaj.081869.
14. de Jonge A, van der Goes B, Ravelli A, et al. Perinatal Mortality and Morbidity in a Nationwide Cohort of 529,688 Low-Risk Planned Home and Hospital Births." *BJOG.* 2009;116 (9): 1177–84. doi:10.1111/j.1471-0528.2009.02175.x.

15. Wiegers T, Keirse M, van der Zee J, Berghs G. Outcome of Planned Home and Planned Hospital Births in Low Risk Pregnancies: Prospective Study in Midwifery Practices in The Netherlands. *BMJ (Clinical Research Ed.)* 1996;313 (7068): 1309–13.
16. Sullivan DA, Weitz R. Labor Pains: Modern Midwives and Home. New Haven/London: Yale University Press;1988:68-80
17. Kennedy H. A model of exemplary midwifery practice: results of a Delphi study. *J Midwifery Womens Health*. 2000;45(4):4-19.
18. North American Registry of Midwives. What is a CPM? www.narm.org/
19. Home Birth Consensus Summit, 2011. What Was the Process Engaging in dialogue: The Future Search model. http://homebirth summit.org/history/what-was-the-process. Accessed July 4, 2012
20. Home Birth Consensus Summit, 2011. Common Ground. http://www.homebirthsummit.org/outcomes/common-ground-statements , Accessed July 4, 2012
21. American Public Health Association. Increasing Access to Out-of-Hospital Maternity Care Services through State-Regulated and Nationally-Certified Direct-Entry Midwives. January 1, 2001. Policy Number: 20013. http://www.apha.org/advocacy/policy/policysearch/default.htm?id=242
22. Dower C, Miller J, O'Neil E and the Taskforce on Midwifery. Charting a Course for the 21st Century: The Future of Midwifery. San Francisco, CA: Pew Health Professions Commission and the UCSF Center for the Health Professions. April 1999. http://www.futurehealth.ucsf.edu/Content/29/1999-04_Charting_a_Course_%20for_the_21st_Century_The_Future_of_Midwifery.pdf
23. Midwives Alliance of North America. Core Competencies. https://mana.org/resources/core-competencies
24. Midwifery Education Accreditation Council. About MEAC. http://meacschools.org/about-meac/
25. Midwives Alliance of North America. Direct-Entry Midwifery State-by-State Legal Status. May 11, 2011. http://www.mana.org/statechart.html

26. NARM North American Registry of Midwives. State Licensure of Certified Professional Midwives. April, 2012. http://narm.org/wp-content/uploads/2012/05/State-Licensure-of-CPMs2012.pdf
27. Schroeder E, Petrou S, Patel N, et al. Cost Effectiveness of Alternative Planned Places of Birth in Woman at Low Risk of Complications: Evidence from the Birthplace in England National Prospective Cohort Study. *BMJ (Clinical Research Ed.)* 2012;344: e2292.
28. Janssen P, Mitton C, Aghajanian J. Costs of Planned Home vs. Hospital Birth in British Columbia Attended by Registered Midwives and Physicians. *PloS One* 2015;10 (7): e0133524. doi:10.1371/journal.pone.0133524.
29. Johnson K, Daviss B. Outcomes of Planned Home Births with Certified Professional Midwives: Large Prospective Study in North America. *BMJ (Clinical Research Ed.)* 2005;330 (7505): 1416. doi:10.1136/bmj.330.7505.1416.
30. Cheyney M, Bovbjerg M, Everson C, Gordon W, Hannibal D, Vedam S. Development and validation of a national data registry for midwife-led births: the Midwives Alliance of North America Statistics Project 2.0 dataset. *J Midwifery Womens Health*. 2014 Jan-Feb;59(1):8-16. doi: 10.1111/jmwh.12165.
31. Cox K, Bovbjerg M, Cheyney M, Leeman L. Planned Home VBAC in the United States, 2004-2009: Outcomes, Maternity Care Practices, and Implications for Shared Decision Making. *Birth* Berkeley Calif. 2015 Dec;42(4):299–308.
32. Bovbjerg M, Cheyney M, Everson C. Maternal and Newborn Outcomes Following Waterbirth: The Midwives Alliance of North America Statistics Project, 2004 to 2009 Cohort. *J Midwifery Womens Health*. 2016 Feb;61(1):11–20.
33. Sandall J, Soltani H, Gates S, Shennan A, Devane D. Midwife-led continuity models versus other models of care for childbearing women. Cochrane Database Syst Rev. 2016 Apr 28;4:CD004667.
34. Molnar S, ICTC, ICAN, MANA, Elephant Circle. Racial Disparities in Birth Outcomes and Racial Discrimination as an Independent Risk Factor Affecting Maternal, Infant, and Child

Health: An Executive Summary of Existing Research. Midwives Alliance of North America; 2015.

35. Cheyney M, Everson C, Burcher P. Homebirth Transfers in the United States: Narratives of Risk, Fear, and Mutual Accommodation. *Qualitative Health Research* 2014;24 (4): 443–56. doi:10.1177/1049732314524028.
36. Allied Midwifery Organizations. Midwifery Education Accreditation Council Website. http://meacschools.org/alliedmidwifery-organizations/.

Massage Therapy

Third Edition (2017) Editors: Stan Dawson, DC, LMBT;
Angie Myer, MA, HHP; Cherie Sohnen-Moe, BA;
Kate Zulaski, BA, BCTMB

Second Edition (2013) Editors: Jan Schwartz, MA, BCTMB;
Stan Dawson, DC, LMBT; Pete Whitridge, BA, LMT

First Edition (2009) Authors: Jan Schwartz, MA, BCTMB;
Cherie Monterastelli, RN, MS, LMT

Second Edition (2013) and Third Edition (2017) Partner
Organization: Alliance for Massage Therapy Education (AFMTE)

First Edition (2009) Partner Organization:
American Massage Therapy Association –Council of Schools (AMTA-COS)

About the Authors/Editors: Schwartz is past Chair of the Commission on Massage Therapy Accreditation (COMTA), co-owner of Education and Training Solutions, LLC, is a member of the ACIH Board and is past Co-chair of the ACIH Education Working Group. Monterastelli previously served as a board member of the American Massage Therapy Association – Council of Schools and ACIH. Dawson is an ACIH Board member, a member of the ACIH Education Working Group, and Vice President of Alliance for Massage Therapy Education (AFMTE). Whitridge is the Immediate Past President of AMTA and former Chair of the Florida Board of Massage Therapy. Myer is the Director of Accreditation for COMTA. Sohnen-Moe is President of AFMTE. Zulaski is the Executive Director of COMTA.

Philosophy, Mission, Goals

Several organizations and scholars have created definitions that incorporate the mission and philosophy of massage therapy (MT). For example, in the frequently used textbook, *Tappan's Handbook of Healing Massage Techniques,* the definition is: "Massage is the intentional and systematic manipulation of the soft tissues of the body to enhance health and healing."[1(p4)] Similarly, according to the American Massage Therapy Association (AMTA) Glossary of Terms, massage therapy is "a profession…with the intention of positively affecting the health and well-being of the client through a variety of touch techniques."

Because most definitions of massage therapy include the use of touch, which is a basic, non-technological approach to health and healing, it follows that most massage therapists subscribe to a natural healing philosophy. That philosophy encompasses a preference for "natural methods of healing, the belief in an innate healing force, and a holistic view of human life."[1(p14)]

With the advent of the Patient Protection and Affordable Care Act of 2010, the healthcare system shows signs of shifting toward collaboration between conventional and complementary healthcare professions, with the goal of wellness in addition to treating illness, and physicians' pay based on health outcomes. With doctor behavior incentivized by payment for keeping people healthy rather than performing more tests and procedures, massage can be particularly well suited to this shift in philosophy. The focus of most massage therapists on health improvement and a positive patient experience, combined with the relatively low cost of massage treatment is aligned with the concerns of the Triple Aim framework developed by the Institute of Healthcare Improvement.

History of the profession

Massage has been practiced in most cultures in both the East and the West throughout human history. Traditional peoples used a variety of techniques that are now known as massage, all of which included some form of person-to-person touching with the intention of manipulating and relaxing the muscles of the body. From the South Sea Islands to the Mexican peninsula and the indigenous cultures of the Americas to the ancient civilizations of Greece, Asia and Africa, historians and researchers find evidence of the practice of massage. In India, the ancient practice of Ayurveda included forms of movement therapy and massage. In Greece, Hippocrates wrote about the ability of massage to build muscle as well as heal it. China and Japan each developed varieties of natural healing that included touch.

The more modern practice of massage, known as Western massage or Swedish massage, became prominent in the 19th and early 20th centuries. Called the father of Swedish massage, Per Henrik Ling (1776-1839) developed a series of exercises that became known as "medical gymnastics" and used a series of movements that applied resistance to the joints. These techniques, however, have little resemblance to massage as it is known today. While there was a hands-on relationship between therapist and subject, the activities were more like those used by physical therapists than massage therapists. Ling's perspective on the practices he was developing evolved, however, and he began to consider the relationship between the physical and mental aspects of wellness, the mind/body connection, which is today very much a part of Swedish massage and other massage modalities.

Not quite a century later, a Dutch physician named Johann Georg Mezger (1838-1909) developed the techniques that are now the basis for Swedish massage:

- Effleurage: Long, gliding strokes
- Petrissage: Lifting and kneading the muscles
- Friction: Firm, deep, circular rubbing movements
- Tapotement: Brisk tapping or percussive movements
- Vibration: Rapidly shaking or vibrating specific muscles

Researchers noticed the similarities between the work of Ling and Mezger and gave Sweden the credit for developing these techniques.

Massage technique has evolved considerably in the past fifty years as therapists strive to be more helpful to their clients and as massage research has advanced. Research and technique advancement is focused primarily on the effects of massage and bodywork on circulation, lymph fluid circulation, muscle tone, fascial patterns, joint neurology, neuroendocrine effects, craniosacral fluid flow and stages of rehabilitation. Pain research projects and the effects of massage on depression and immune function show considerable promise.

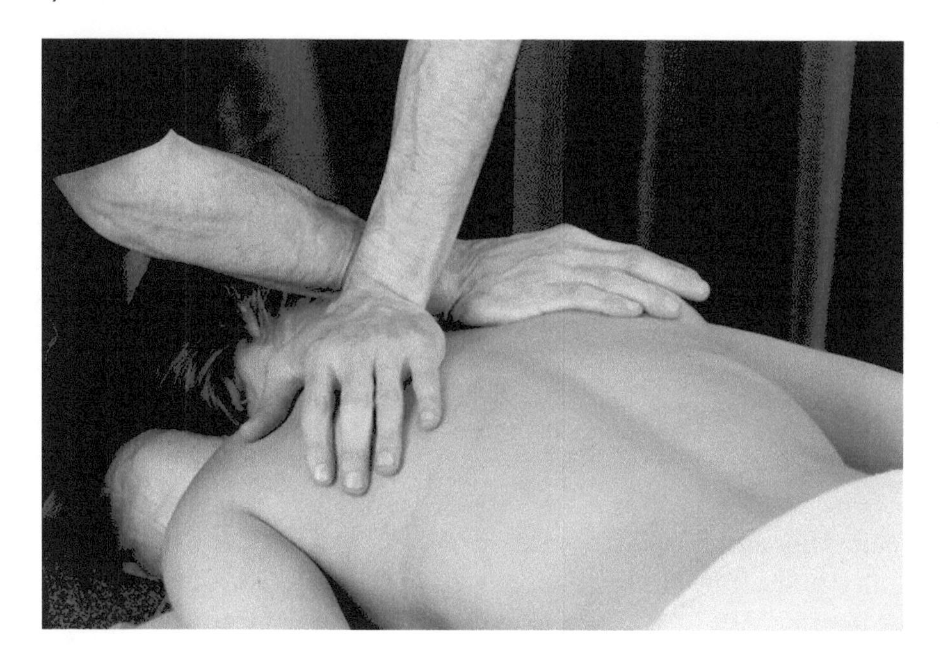

Characteristics and Data*

Industry research estimates that there are 300,000-350,000 practicing massage therapists and massage school students in the US. Today's massage therapists are likely to enter the massage therapy profession as a second career and are predominantly female (86%). The median age for massage therapists is 45 years old, with 21% younger than 35.

Massage therapists earn a comparable annual income to other healthcare support workers, according to the US Department of Labor Statistics. In 2015, the average annual income for a massage therapist was $38,040. A 2016 AMTA survey indicates that the annual income for a massage therapist is $24,519. Forty-five percent (45%) reported also earning income working in another profession. This is down from 53% in 2012. The majority of massage therapists are self-employed. They work an average of 20 hours a week up from 15 hours a week in 2012.

---

* Data from the American Massage Therapy Association Industry Fact Sheet, February, 2016, unless otherwise noted. https://www.amtamassage.org/infocenter/economic_industry-fact-sheet.html.

Growth in the healthcare industry is providing numerous jobs for massage therapists. There has been an increase in the number of massage therapists directly employed by spas and clinics or employed or contracted to work in healthcare settings. From 2005 to 2015, the percentage of massage therapists who worked in a healthcare environment increased from 10% to 23%. Of the adults who received massage between July 2014 – July 2015, 52% report doing so for "medical or health reasons." The AMTA consumer survey found that 51 million adults had discussed massage therapy with their doctors or healthcare providers, and 69% of those doctors either referred, strongly recommended, or encouraged them to get a massage. This has resulted in 55% of massage therapists reporting they have received a referral from a hospital or medical office in the last 6 months, with 11% indicating they receive referrals at least once a week.

Clinical Care

Approach to patient care

Massage therapists in clinical settings typically perform an intake interview focusing on the client's health history, chief complaints and main reason for seeking care. After the intake, a massage therapist usually performs an assessment in order to determine contraindications, need for referral, appropriateness of treating and a plan for a treatment session.

During the treatment session, the massage therapist employs a variety of modalities intended to influence blood flow, lymph flow, muscle tone, fascial length and organization, joint dynamics, joint neurology, and/or craniosacral fluid flow dynamics. Treatments in a clinical setting are typically structured into a treatment plan either by the massage therapist or the health professional the massage therapists' work is supporting or complementing.

Swedish massage (Western Massage) is designed to facilitate relaxation, stress relief through parasympathetic dominance. Advanced level massage can support pain and symptom relief, structural/postural balancing and wellness-oriented collaboration with the entire healthcare team. Advanced level massage therapists may also have specialized training in lymphatic drainage, myofascial manipulation, neuromuscular therapeutic techniques, visceral manipulation, infant massage, pregnancy massage, oncology massage, geriatric massage, sports massage, Asian bodywork therapies or other specialties in massage and bodywork.

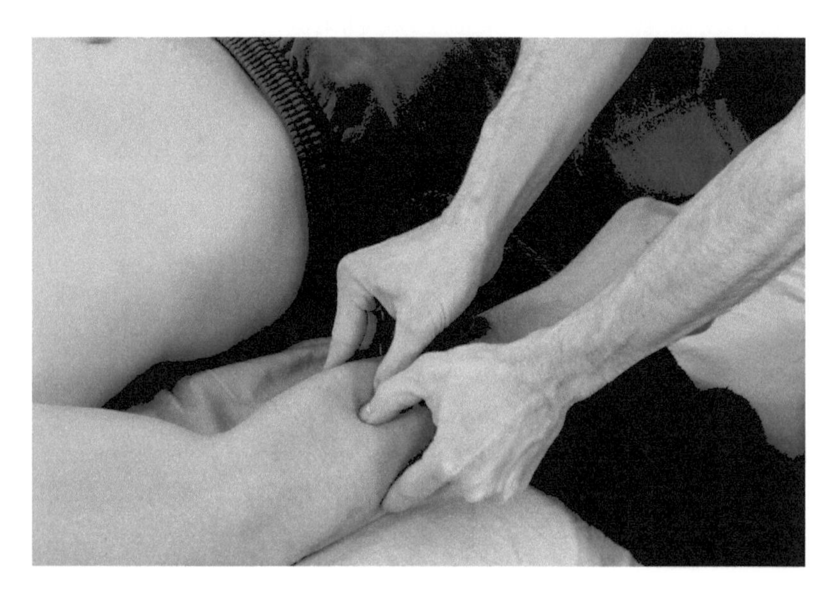

Scope of practice

The scope of practice and common terms are similar in most of the state regulations, but there are significant differences between states. All describe massage as soft tissue manipulation to improve health or some similar idea. All exclude the authority to make a diagnosis, prescribe medications, perform chiropractic adjustments, offer injections or venipuncture, or practice acupuncture, or psychotherapy. Entry level training prepares practitioners for a part of the potential scope of practice afforded massage therapists. Practitioners are *ethically* obligated to limit their practice to what they have been adequately trained to do. Post-graduate training in Specialty Certifications and continued education courses prepare practitioners in areas not covered in entry level training. This advanced training enables practitioners to utilize more of the techniques allowed by state regulation that are not covered in entry level training.

There are overlapping scopes of practice with chiropractic and physical therapy. Both professions are allowed to perform massage under their respective scopes of practice. The scope of practice for acupuncture and Oriental Medicine allows their practitioners to perform Asian bodywork, which can appear similar to massage while focused on the energy system rather than the neuromusculoskeletal system.

Referral practices

Most massage therapists are adequately trained to know how and when to refer to other healthcare professionals. Because of inconsistencies in training and experience, and the broad range of educational requirements noted below, practitioners' skills in the area of referral practices can vary. Advanced practitioners, however, appreciating the interdependence of mental, emotional, musculoskeletal and visceral elements of the spirit-mind-body system, value collaboration with both conventional and other complementary healthcare professionals. Mature professionals understand the appropriateness of helping their clients find care that is efficacious, cost effective and patient-centered. This leads to an attitude of comfort with referral to the appropriate healthcare practitioner whenever indicated.

Since massage therapists do not diagnose, they need to confer on cases with physicians and other healthcare practitioners in order to judge case management issues, when indicated. Massage therapists are trained to refer to lists of possible contraindications for which referral may be required or advised.

Third-party payers

Some third parties reimburse for massage therapy, however, massage therapy is not covered in Centers for Medicare and Medicaid Services legislation. Payments are most often made for workers' compensation claims and auto accident injuries. Additional reimbursements vary by state and insurance provider. The state of Washington, for instance, mandates that all health plans include every category of licensed provider in their benefit design. Massage therapists are receiving an increasing number of referrals for healthcare treatments.

The AMTA continues to strategically collaborate with associations and organizations who share a vision for the importance of massage therapy in one's health and wellness regimen. As such, AMTA is sustaining its representation to the American Medical Association's (AMA's) Current Procedural Terminology (CPT) Health Care Professional's Advisory Committee. AMTA's representatives provided input into decisions related to CPT code use by massage therapists. This is of direct benefit to massage therapists who seek insurance reimbursement.

Integration Activities

A number of partnerships have developed between massage schools and medical schools. These partnerships can be as minimal as medical schools referring students to massage schools in the local area or as involved as jointly developing a multidisciplinary student clinic or a curriculum for use in each other's institutions.

Massage therapists are working in many integrated settings. These settings include hospitals, wellness centers, chiropractic offices, acupuncture and Oriental Medicine offices, sports medicine clinics, physical therapy clinics, and holistic health centers.

Massage therapists working with ACIH have helped in the development of the ACIH Competencies for Optimal Practice in Integrated Environments which the Alliance for Massage Therapy Education has endorsed. Massage therapists working with ACIH have also been involved in interprofessional education (IPE) conferences and research conferences. If the massage profession adopts and prepares massage therapists in these competencies, massage will be positioned to enter the healthcare system of the future with its focus on integrated care.

Education

Schools and programs

According to the Associated Bodywork and Massage Professionals (ABMP) 2014 enrollment survey, there were 1,274 state-approved massage therapy training schools or programs in the US. Of those schools/programs, 57% were accredited by agencies recognized by the US Department of Education. Sixty-five of those accredited schools/programs are Commission on Massage Therapy Accreditation (COMTA) accredited schools/programs. COMTA is the only specialized accreditor for massage/bodywork and esthetics.[2]

The AMTA industry survey reports massage therapists have an average of 671 hours of initial training. In most cases, the total number of hours a school offers trends closely to what is required by that state for licensure. This varies from state to state. The minimum number of hours required by COMTA is 600.

The Alliance for Massage Therapy Education was organized in 2009 as a 501(c)(6) non-profit corporation. It was created to "fill a gap" as the profession was missing a not-for-profit dedicated to education. There are three member groups represented by the Alliance: schools, continuing education providers, and teachers. The AFMTE was designated by the ACIH board of directors to represent massage educators in October 2011. The AFMTE is dedicated to improving the quality of massage therapy and bodywork education by raising teacher standards, curriculum standards and education standards. The AFMTE created a set of core competencies for massage therapy and bodywork teachers in 2012 to be used as the basis for a national teacher certification program and for implementation in accreditation standards and state regulations to bolster the quality of massage education. They also hold a biennial Educational Congress focused on fostering collaboration in the massage therapy and bodywork education community.

A national membership organization, a national business with a membership component, and several state organizations provide services and support to schools. The two national groups are the not-for-profit American Massage Therapy Association (AMTA) and the for-profit Associated Bodywork and Massage Professionals (ABMP).

AMTA offers school memberships that include benefits to help administrators recruit students and operate successful schools as well as benefits to support teachers with classroom resources. Education offerings include a series of online courses for teachers, a teacher track at the AMTA National Convention as well as an annual Schools Summit where teachers from around the country convene for practical advice and best practice sharing. School members receive optional liability insurance and are listed in AMTA's FindaMassageSchool.org locator service promoted to thousands of prospects each month. Students at AMTA member schools receive free AMTA Student membership and benefit from instructional resources like the free Massage Exam Study App. They also have complimentary access to AMTA's resume builder and Job Bank where they can post their resume, peruse job postings, and sign up for job alerts. AMTA has long been an advocate of fair and consistent licensing of massage therapy. The Government Relations Department works with schools and chapters to develop regulations that support high standards for the profession.

The ABMP has a staff of six professionals who work directly with the schools. School memberships include materials and resources for students and faculty, classified ads and job postings, an e-mail account and online networking forums, optional school liability and business property insurance. Most importantly, the ABMP offers an annual ABMP School Issues Forum for schools to network and discuss common challenges. The ABMP has an extensive website with archived webinars, teaching tips, learning checklists, and rubrics to support faculty and student success. There is also a regional program known as *Instructors on the Front Lines*. These are free one-day seminars to support professional teacher development.

Curriculum content

Unlike the other disciplines within ACIH, there are neither required national standards nor a single body that the entire massage profession has designated to determine the knowledge and skills required to call one a member of the field. There is a specialized accrediting agency for massage therapy which is recognized by the US Department of Education, the Commission on Massage Therapy Accreditation (COMTA), which developed national curriculum competencies to meet the diverse needs of the profession. Graduation from an accredited school is not universally required for licensure. Fewer than 10% of massage programs are currently accredited by COMTA and beholden to the national competencies. The majority of schools/programs match their curriculum to state licensing, which can range from 500 hours – 1,000 hours. There exists, therefore, a wide variety in the type and quality of education available in massage schools and programs.

In 2014, the Entry-Level Project Analysis (ELAP) was released and endorsed by all major massage therapy organizations. The ELAP recommends a minimum of 625 hours for entry-level training and provides a detailed blueprint of learning outcomes which should be included. This document has been widely circulated, but it is not yet clear the impact it will have on school curriculum adoption or licensing requirements.

Many practitioners choose to take continuing education courses to gain more extensive training in a specific modality or area of practice. Requirements for continuing education vary from state to state depending on state licensure requirements, according to the membership organization a practitioner chooses to join, and depending on whether the practitioner elects to maintain the national certification credential.

Until national curriculum standards are universally adopted by the profession, state regulations, certification and accreditation standards dictate curriculum content to the schools. This arrangement is problematical for the maturation of the profession and for public health clarity because the state curriculum requirements are neither rigorous nor consistent, and almost 50% of the schools are not subject to review by any accreditation agency, yet all are allowed to register as "approved" schools.

Accreditation

The Commission on Massage Therapy Accreditation (COMTA) is recognized by the US Department of Education as specializing in massage. COMTA was started by the massage profession in 1982 and granted recognition by the US Department of Education in 2002. COMTA provides either institutional or programmatic accreditation. Both designations include a comprehensive evaluation of curriculum competencies which outline what graduates must know and be able to do. Institutional accreditation is available to schools teaching only massage, bodywork and/or esthetics. Programmatic accreditation is available to massage therapy and bodywork programs within schools which already have institutional accreditation from another recognized agency.

Accreditation by COMTA requires that at least one program in massage or bodywork offered by the school contains 600 clock hours of content and includes teaching and assessment of the required competencies. The COMTA competencies outline requirements in six primary areas:

1. Plan and organize an effective massage and bodywork session (includes sciences)
2. Perform massage therapy and bodywork for therapeutic benefit
3. Develop and implement a self-care strategy
4. Develop a successful and ethical therapeutic relationship with clients
5. Develop a strategy for a successful practice, business or employment situation
6. Identify strategies for professional development.

In 2016, the Commission introduced an alternative recognition status called COMTA-Endorsed Curriculum, which is not accreditation but does evaluate the curriculum for meeting the curriculum competencies.

Regulation and Certification

Regulatory status

Forty-six states (plus the District of Columbia, Puerto Rico and US Virgin Islands) have some form of regulation in place for massage therapy. There is a lack of consistency among these licensure laws noted earlier, with requirements ranging from 500 hours – 1,000 hours and different content requirements. Depending on the state law, massage therapists can be referred to as licensed, state certified, or registered. In most cases, only individuals who have the state designation may perform massage or use a title indicating that they perform massage.

In 2014, the Federation of State Massage Therapy Boards (FSMTB) published a Model Practice Act, a recommendation to state licensing boards around the country on what should be included in future laws. Among other details, the Model Practice Act recommends that "all educational institutions adopt a curriculum that reflects the ELAP recommendations [625 hours] and that is acceptable to an accrediting body recognized by the US Department of Education." As noted earlier, it has yet to be determined the impact of this recommendation.

Examinations and certifications

The Federation of State Massage Therapy Boards (FSMTB), established in 2005, is currently made up of 43 state licensing boards and agencies that regulate the massage therapy and bodywork profession. The mission of the FSMTB is to support its member boards in their work to ensure that the practice of massage therapy is provided to the public in a safe and effective manner. In October 2007, FSMTB created an entry-level national licensure examination to serve the needs of the regulatory community in licensing massage therapists after completing a Job Task Analysis (JTA). Currently, 44 states accept the FSMTB exam. For detailed results and a state-by-state compendium of laws, please visit the FSMTB website listed in the Resources section.

The National Certification Board for Therapeutic Massage and Bodywork (NCBTMB) is an independent, private, nonprofit organization that was founded in 1992 to establish a certification program and uphold a national standard of excellence. NCBTMB's exam programs are accredited by the Institute for Credentialing Excellence (ICE). In 2013, NCBTMB's national certification transitioned to a board certification in massage therapy. This credential requires a minimum of 750 hours of training, 250 hours of documented hands-on work experience, and passage of NCBTMB's exam. The exam is designed to measure critical thinking rather than simple recall. Additionally, the program includes adherence to the NCBTMB Code of Ethics and Standards of Practice, passing a national background check along with other eligibility criteria. The last component of this credential is a commitment to lifelong learning and re-certification. This new credential is the first of its kind for the massage profession, offering a true separation between licensing and certification. Since 2013, the NCBTMB has also worked to include a variety of specializations for practitioners to advance their skills and credentials in particular fields, such as Integrative Health, Sports, and Military Veteran Massage.

Research

The Massage Therapy Foundation's (MTF) mission is to advance the knowledge and practice of massage therapy by supporting scientific research, education, and community service. The Foundation funds research studies investigating the many beneficial applications of massage therapy. Foundation research grants are awarded to individuals or teams conducting studies that promise to advance the understanding of specific therapeutic applications of massage, public perceptions of and attitudes toward massage therapy, and the role of massage therapy in healthcare delivery.

The Massage Therapy Foundation commissioned a research agenda in 1999, recommending areas of research in massage therapy and bodywork that are most needed. This agenda has been under review since 2013. Investigators who apply for MTF research grants are referred to the agenda and are encouraged to address one of the following in their research:

- Build a massage research infrastructure
- Fund studies into safety and efficacy
- Fund studies of physiological (or other) mechanisms (how massage works)
- Fund studies stemming from a wellness paradigm

- Fund studies into the profession of therapeutic massage and bodywork

The MTF launched the *International Journal of Therapeutic Massage and Bodywork: Research, Education and Practice* (IJTMB) in August, 2008. It is a free, online, peer-reviewed journal, which is also catalogued in PubMed.

In addition to its efficacy for muscle and other soft tissue ailments, research has shown that massage therapy can relieve symptoms associated with many health issues, such as osteorarthritis.[3] Among the other ways massage therapy has been shown to be effective are:

- **Relief of back pain**
  More than 100 million Americans suffer from low-back pain, and nearly $25 billion a year is spent in search of relief. A 2003 study showed that massage therapy produced better results and reduced the need for painkillers by 36% when compared to other therapies. Today, massage therapy is one of the most common ways people ease back pain.[4]
- **Treating Migraines**
  Of the 45 million Americans who suffer from chronic headaches, more than 60% suffer from migraines. For many, it's a distressing disorder that is triggered by stress and poor sleep. In a recent study, massage therapy recipients exhibited fewer migraines and better sleep quality during the weeks they received massage, and the three weeks following, than did participants who did not receive massage therapy. Another study found that in adults with migraine headaches, massage therapy decreased the occurrence of headaches, sleep disturbances, and distress symptoms. It also increased serotonin levels, believed to play an important role in the regulation of mood, sleep, and appetite.[5,6]
- **Easing Symptoms of Carpal Tunnel**
  Carpal tunnel syndrome is a progressively painful condition that causes numbness and tingling in the thumb and middle fingers. Traditional treatments for carpal tunnel range from a wrist brace to surgery. However, a 2004 study found that carpal tunnel

patients receiving massage reported significantly less pain, fewer symptoms, and improved grip strength compared to patients who did not receive massage.[7]

- **Reducing Anxiety**

  An estimated 20 million Americans suffer from depression. A review of more than a dozen massage studies concluded that massage therapy helps relieve depression and anxiety by affecting the body's biochemistry. In the studies reviewed, researchers measured the stress hormone cortisol in participants before and immediately after massage and found that the therapy lowered levels by up to 53%. Massage also increased serotonin and dopamine, and neurotransmitters that help reduce depression.[8]

- **Alleviating Symptoms and Side Effects of Cancer**

  Massage therapy is increasingly being applied to symptoms experienced by cancer patients, such as nausea, pain, and fatigue. Researchers at Memorial Sloan-Kettering Cancer Center asked patients to report the severity of their symptoms before and after receiving massage therapy. Patients reported reduced levels of anxiety, pain, fatigue, depression, and nausea, even up to two days later.[9] In a study of breast cancer patients, researchers found that those who were massaged three times a week reported lower levels of depression, anxiety and anger, while increasing natural killer cells and lymphocytes that help to battle cancerous tumors.[10,11]

- **Lowering Blood Pressure**

  Hypertension, if left unchecked, can lead to organ damage. Preliminary research shows that hypertensive patients who received three 10-minute back massages a week had a reduction in blood pressure, compared to patients who simply relaxed without a massage.[12]

Challenges and Opportunities

The following descriptions are intended to assist the reader in understanding aspects of the massage therapy profession rather than formal positions taken by the massage field on the issues.

Key challenges 2016-2020

- The lack of consistent educational quality between massage schools and programs, and the resulting skill discrepancies in practitioners negatively impacts the profession in many ways, from practitioner employability or salaries, to reliable research, to acceptance into integrative healthcare. The Entry Level Analysis Project (ELAP) and the Massage Therapy Body of Knowledge (MTBOK) before it were steps toward a solution, but since these are not regulatory documents, more is needed for consistent implementation.
- Massage therapy educational programs need to move to greater standardization in the content, scope, and length of entry-level training programs. This could be accomplished by revising educational standards based on the MTBOK, COMTA curriculum standards and the ELAP.
- Faculty qualifications vary greatly from school to school with no commonly agreed upon formal training for massage therapy and bodywork instructors. Such training is also part of creating consistency in educational programs for massage therapy and bodywork. Without consistency in educational programs and faculty qualifications, consistency in terminology will prove difficult. Consumer expectations will also be affected. Adopting the AFMTE Teacher Standards and basing teacher training on them could be significant steps toward accomplishing this.
- Now that more states are regulated, the lack of portability of massage licensure is a growing concern. It is cumbersome, and in some case not possible, to move to another state and establish licensure in the new state without starting training over from the beginning if their school wasn't approved in the new state. A

common language is needed, not only for the number of hours for each state, but the curriculum content criteria and the process of verifying educational credentials.

- The number of massage schools and students enrolled in these schools have contracted as practitioners struggle to make a living, employers struggle to find qualified employees. Ironically, this is happening at a time when consumer demand for massage and acceptance into healthcare appears to be increasing, which begs for collaboration and solutions to capitalize on the opportunities.

Key opportunities 2016-2020

- Further utilization and promotion of COMTA programmatic accreditation to assure consistent curriculum content, educational quality and faculty qualifications. Massage therapy remains one of the few healthcare professions without mandatory accreditation requirements. Requiring accreditation has the potential to improve massage therapy's credibility in integrative healthcare. Further, as the only national massage regulatory body, COMTA is uniquely positioned to assist with implementation of industry recommendations to consistently evaluate educational quality and facilitate license portability.
- Further development of industry partnerships with potential massage therapy employers will increase knowledge of massage therapy training and credentials among those industries, and provide more opportunities for collaboration in client care. Partnership with employers will, in turn, give the massage education institutions much needed information on what employers' needs are so curriculum can be developed to meet those needs and enhance the likelihood of job placement for graduates.
- Development of the profession's research base will allow practitioners to provide more effective, outcome-based massage therapy for their clients, which will provide better results and increase the demand for massage therapy. Establishing awareness of and interest in massage as an effective therapy will increase acceptance by healthcare providers and third-party payers.

Greater communication within the profession will also increase as a standardized scientific language is more commonly used.

- As the AFMTE completes the National Teacher Education Standards Project (see Resources section below), accrediting agencies and state licensing boards could adopt the requirement that massage and bodywork educators be certified. Until that occurs, teachers and schools could voluntarily seek teacher certification and schools could distinguish themselves by requiring their faculty to be certified massage and bodywork educators once the NTESP is completed.
- As massage therapy becomes a more integral component of the complementary and integrative aspects of health care, there will be an increasing demand for quality, standardization, and research. This demand will help set the parameters for the profession to evolve into a more recognized, valuable, and respected part of health care.
- Improvement of industry partnerships with all healthcare professions, both conventional and complementary and integrative.
- Implementation of the Model Practice Act to bring higher standards and consistency between state licensing boards, as well and facilitate license portability for practitioners.

Resources

Organizations and websites

The profession is represented by seven organizations which meet yearly as a coalition:

- Alliance for Massage Therapy Education (AFMTE) was founded in 2009 as a nonprofit organization to serve as an independent voice, advocate, and resource for the quality of massage and bodywork education. AFMTE is comprised of schools, faculty members, and continuing educators in massage and bodywork. The Alliance is working to advance massage education through its National Teacher Education Standards Project (NTESP). The

NTESP includes a set of competencies for massage educators and a teacher certification program, still in development, based on this set of competencies. The concept is to improve massage and bodywork education by requiring massage and bodywork educators to be trained both as educators and as content experts. AFMTE develops and advocates for improved standards of education for the massage profession. AFMTE represents the council of colleges/schools for massage therapy within ACIH.

- American Massage Therapy Association (AMTA) is a not-for-profit membership organization that represents more than 500 massage therapy schools and programs and has more than 70,000 member massage therapists. AMTA works to establish massage therapy as integral to the maintenance of good health and complementary to other therapeutic processes, and to advance the profession by promoting certification and school accreditation, ethics and standards, continuing education, professional publications, legislative efforts, public education, and fostering the development of members. AMTA offers liability insurance to its members.
- Associated Bodywork and Massage Professionals (ABMP) is a private business with professional membership dimensions founded in 1987 to provide massage and bodywork practitioners with professional services, information, and public and regulatory advocacy. ABMP has initiatives in the areas of promoting ethical practices, protecting the rights of practitioners, and educating the public regarding the benefits of massage and bodywork. Its current membership totals over 82,000. Members must adhere to a published code of ethics.
- Commission on Massage Therapy Accreditation (COMTA) is a non-profit accrediting agency recognized by the US Department of Education specializing in massage therapy/bodywork and esthetics/skin care. It was founded as part of the AMTA Council of Schools in 1982 but is now fully independent. COMTA offers both institutional and programmatic accreditation, as well as a new recognition status of COMTA-Endorsed Curriculum.

- Federation of State Massage Therapy Boards (FSMTB) was formed in 2005 to bring the regulatory community together and provide a forum for the exchange of information. The result of the exchange was the development of a licensure exam which was introduced in 2007. The FSMTB's mission is to support its member boards in their work to ensure that the practice of massage therapy is provided to the public in a safe and effective manner. There are currently 43 member boards with 44 states and territories accepting the exam for licensing.
- Massage Therapy Foundation was founded by the American Massage Therapy Association in 1990 with the mission of advancing the knowledge and practice of massage therapy by supporting scientific research, education, and community service. The Foundation is able to provide these services by individual gifts, industry support, and fundraising events.
- National Certification Board for Therapeutic Massage and Bodywork (NCBTMB) was founded in 1992 as a nonprofit organization to establish a national certification program and uphold a national standard of professionalism. NCBTMB works to foster high standards of ethical and professional practice through a recognized, credible credentialing program that assures the competency of practitioners of therapeutic massage and bodywork. NCBTMB also has outreach programs for stakeholders, including schools and state boards. The organization had certified more than 90,000 massage therapists and bodyworkers as of the date of publication of this booklet. NCBTMB's certification program is made up of a number of components: eligibility, examinations, code of ethics, standards of practice, continuing education and recertification. NCBTMB's certification program is accredited by the National Commission for Certifying Agencies (NCCA), the accrediting branch of the National Organization for Competency Assurance (NOCA).

- Alliance for Massage Therapy Education (AFMTE)
  www.afmte.org
- American Massage Therapy Association (AMTA)

www.amtamassage.org
- Associated Bodywork and Massage Professionals (ABMP)
  www.abmp.com
- Commission on Massage Therapy Accreditation (COMTA)
  www.comta.org
- Federation of State Massage Therapy Boards (FSMTB)
  www.fsmtb.org
- Massage Therapy Foundation
  www.massagetherapyfoundation.org
- National Certification Board for Therapeutic Massage and Bodywork (NCBTMB)
  www.ncbtmb.org

Bibliography

Benjamin BE, Sohnen-Moe C. *The Ethics of Touch*. Tucson, Arizona: Sohnen-Moe Associates; 2014.

Biel A. *Trail Guide to the Body*. 4th ed. Boulder, CO: Books of Discovery; 2010.

Cohen B. *Memmler's the Human Body in Health and Disease* (12th ed.). Philadelphia: Lippincott, Williams & Wilkins; 2008.

Dryden T, Moyer C, eds.. *Massage Therapy: Integrating Research and Practice*. Champaign, IL: Human Kinetics; 2012.

Fritz S. *Fundamentals of Therapeutic Massage* 4th ed. St. Louis, Missouri: Mosby; 2009.

Frye B. *Body Mechanics for Manual Therapists: A Functional Approach to Self-care.* 2nd ed. Stanwood, Washington: Fryetag; 2004.

Hymel GM. *ResearchMmethods for Massage and Holistic Therapies*. St. Louis, Missouri: Mosby; 2006.

Lowe W. *Orthopedic Massage: Theory and Technique*. St. Louis, Missouri: Elsevier Health Services; 2003.

Menard MB. *Making Sense of Research: A Guide to Research Literacy for Complementary Practitioners*. Toronto, Ontario, Canada: Curties-Overzet; 2009.

Rattray F, Ludwig L. *Clinical Massage Therapy: Understanding, Assessing and Treating over 70 Conditions*. Elora, Ontario, Canada: Talus; 2001.

Sohnen-Moe C.*Business Mastery: A Guide for Creating a Fulfilling, Thriving Practice and Keeping it Successful*. Tucson, AZ: Sohnen-Moe Associates; 2016.

Thompson DL. *Hands Heal: Communication, Documentation, and Insurance Billing for Manual Therapists*.3rd ed. Philadelphia, PA: Lippincott, Williams & Wilkins; 2006.

Werner R. *A Massage Therapist's Guide to Pathology*. 5th ed. Philadelphia, PA: Lippincott, Williams & Wilkins; 2012.

Williams A, ed. *Teaching Massage: Fundamental Principles in Adult Education for Massage Program Instructors*. Philadelphia, PA: Lippincott, Williams & Wilkins; 2008.

Yates J. *A Physician's Guide to Therapeutic Massage*. Toronto, Ontario, Canada: Curties-Overzet; 2004.

Citations

1. Benjamin P, & Tappan F. *Tappan's Handbook of Healing Massage Techniques: Classic, Holistic, and Emerging Methods* 5th ed. Boston, MA: Pearson Education; 2009.
2. Specialized Accreditation. Commission on Massage Therapy Accreditation website. https://comta.org/accreditation/ Accessed April 17, 2017
3. Perlman A, Sabina A, Williams AL, et al. Massage therapy for osteoarthritis of the knee: a randomized trial. Arch Intern Med. 2006;166(22):2533-2538.
4. Cherkin DC, Sherman KJ, Devo RA, et al. A review of the evidence for the effectiveness, safety, and cost of acupuncture, massage therapy, and spinal manipulation for back pain. *Ann Intern Med*. 2003;138(11):898-906.
5. Lawler SP, Cameron LD. A randomized, controlled trial of massage therapy as a treatment for migraine. *Ann Behav Med*. 2006;32(1):50-9.
6. Field T, Hernandez-Reif M, Miguel Diego M, et al. Serotonin and dopamine increase following massage therapy. *Int J Neuroscience*. 2005;115(10):1397-1413.
7. Cambron, J., Dexheimer, J., Coe, P., Swenson, R. Side-effects of massage therapy: a cross sectional study of 100 clients. *J. Comp Alt Med*, 2007; 13(8), 293-796.

8. Cassileth BR, Vickers AJ. Massage therapy for symptom control: outcome study at a major cancer center. *J Pain Symptom Manage.* 2004;28(3):244-9.
9. Hernandez-Reif M, Ironson G, Field T, et al. Breast cancer patients have improved immune and neuroendocrine functions following massage therapy. *J Psychosom Res.* 2004;57(1):45-52.
10. Moraska A, Chandler C. Changes in psychological parameters in patients with tension-type headache following massage therapy: A pilot study. *J Man Manip Ther* 2009;17(2):86-94.
11. Moyer CA, Rounds J & Hannum J. A meta analysis of massage therapy research. *Psychol Bull.* 20004:130(1):3-18.
12. Olney CM. The effect of therapeutic back massage in hypertensive persons: a preliminary study. *Biol Res Nurs.* 2005;7(2):98-105.

Naturopathic Medicine

Third Edition (2017) Editors: Christa Louise, PhD, MS;
Daniel Seitz, JD, EdD; JoAnn Yanez, ND, MPH, CAE

Second Edition (2013) Editors: Christa Louise, PhD, MS;
Paul Mittman, ND, EdD; Michael Traub, ND, DHANP, FABNO;
Marcia Prenguber, ND, FABNO; Elizabeth Pimentel, ND

First Edition (2009) Authors: Paul Mittman, ND, EdD,
Patricia Wolfe, ND, Michael Traub, ND, DHANP, FABNO

Partner Organization: Association of Accredited Naturopathic
Medical Colleges (AANMC)

About the Authors/Editors: Mittman is President and CEO of Southwest College of Naturopathic Medicine and Health Sciences and past President of the Association of Accredited Naturopathic Medical Colleges. Wolfe is the Founding President of the Boucher Institute of Naturopathic Medicine and a past Association of Accredited Naturopathic Medical Colleges member. Traub is a private practitioner and a past President of the American Association of Naturopathic Physicians. Louise is the Executive Director of the North American Board of Naturopathic Examiners and is a former ACIH Board member. Prenguber is Dean of the University of Bridgeport, College of Naturopathic Medicine, is an ACIH Board member and is Co-chair of the ACIH Clinical Working Group. Pimentel is Dean of Naturopathic Medicine at the Maryland University of Integrative Health. Seitz is Executive Director of the Council on Naturopathic Medical Education. Yanez is the Executive Director of the Association of Accredited Naturopathic Medical Colleges and ACIH vice chair.

Philosophy, Mission, and Goals

Although naturopathic medicine as an organized system of practice was first conceived at the beginning of the 20th century, it has its roots in traditions that date back to the time of Hippocrates. Naturopathic medicine is a holistic approach to health care that recognizes and respects the individuality of the patient. The American Association of

Naturopathic Physicians (AANP) defines naturopathic medicine as "a distinct primary health care profession, emphasizing prevention, treatment and optimal health through the use of therapeutic methods and substances which encourage the person's inherent self -healing process, the vis medicatrix naturae."[1]

Naturopathic medicine has its foundation in the biomedical sciences, and is a comprehensive system of health care. Naturopathic physicians act to stimulate the inherent self-healing abilities of the individual through lifestyle change and the application of non-suppressive therapeutic methods and modalities, including clinical nutrition, botanical medicine, homeopathy, physical medicine, and health psychology. Naturopathic physicians are also schooled in and able to apply conventional clinical practices including emergency medicine, minor surgery, pharmacology, and natural childbirth.[2] Naturopathic physicians respect and apply the principles of traditional world practices such as Traditional Chinese Medicine and Ayurvedic medicine, which provide an added basis for understanding the individual patient from a constitutional perspective.

The techniques of naturopathic medicine include traditional and modern, empirical and scientific methods.[3] Naturopathic medicine is distinguished by the principles upon which its practice is based. The application of these principles in practice is continually reexamined in the light of scientific advances.

The philosophy of naturopathic medicine is embodied by six principles:

First, Do No Harm (*primum no nocere*)
Naturopathic physicians strive to minimize harmful side effects, using the least invasive means necessary to diagnose and treat the patient.

Use the Healing Power of Nature (*vis medicatrix naturae*)
Naturopathic physicians recognize and seek to strengthen the inherent ability of the body to heal itself. The naturopathic physician's role is to work with the patient to identify and

remove obstacles to healing, and to facilitate and enhance the self-healing process.

Identify and Treat the Cause
Naturopathic physicians strive to identify and address the underlying causes of disease.

Treat the Whole Person
Naturopathic physicians recognize that the health of the patient must be addressed at all levels: physical, mental, emotional, spiritual, social, and environmental.

Doctor as Teacher (docere)
A primary role of the naturopathic physician is to educate the patient and empower the individual to take responsibility for her/his own health. Naturopathic physicians recognize and foster the therapeutic power of the doctor/patient partnership.

Focus on Prevention
Naturopathic physicians believe that preventing disease is preferable to waiting until the disease must be treated. This focus includes assessing risk factors and susceptibility to disease, encouraging lifestyle choices that prevent diseases from manifesting, and intervening in disease processes to slow their progression.

Naturopathic medicine is aimed at strengthening the body's natural self-regulatory (self-healing) ability. In the naturopathic model, therapies are applied to reestablish physiological and psychological homeodynamics. Symptoms are seen as the body's way of alerting the patient to pay attention to some aspect of her/his health. Naturopathic physicians believe that there are detrimental health consequences to merely suppressing symptoms without addressing the cause. For example, the naturopathic model recognizes fever as the body's attempt to heal itself. Naturopathic treatment might be aimed at stimulating the fever (while monitoring the process to prevent an elevation that would be dangerous), with the knowledge that when the temperature has reached a certain level, the body's self-regulatory system will be activated and the trajectory of the temperature rise will be reversed. The naturopathic model posits that even specific diseases will benefit from non-specific therapies, because these non-specific therapies will enhance the body's ability to self-regulate. The use of homeopathy in naturopathic medicine is exemplary of this assertion that when the body is given the information it needs, it will self-correct.

Several key features of the naturopathic model distinguish it from the allopathic model:

1. The distinction between disease (as a physiological process) and illness (as the psychological experience of the condition) is de-emphasized in the naturopathic model. The term *imbalance* is used to refer to both the biological aspects that initiate or exacerbate the problem and to the patient's experience. Both the biological aspects and the patient's experience must be addressed.
2. The naturopathic model is holistic. It recognizes the interconnectedness of physiological systems (e.g., the central nervous system and the immunological system) and of systems at different levels of the organism (e.g., the physical level and the psychological level). Naturopathic medicine features a multifactorial approach to healing (e.g., a treatment plan that includes specific therapeutics, diet, exercise, and stress management).
3. Naturopathic medicine works within the framework of self-regulation, i.e., it works with the biological systems that have been organized to maintain health. This is apparent in the belief that the body has the power to heal itself and that the role of the physician is to remove obstacles to that healing.
4. Naturopathic medicine works within as well as between levels of the organism. Healing at the deepest levels cannot occur until healing has taken place at the more superficial levels. Symptoms that are suppressed without addressing the cause will later manifest in other ways, potentially at deeper levels which will lead

to more serious disease states. Therefore, healing must be addressed at all levels.

5. Because naturopathic philosophy acknowledges multifactorial etiology to dysfunction, naturopathic treatment is highly individualized, taking into account the unique circumstances of the patient's physiology and life. Treatment is specifically targeted toward the individual patient's particular areas of dysregulation.
6. The patient-physician relationship is seen as a partnership, with the patient collaborating in treatment. This relationship is recognized as being a vital aspect of the healing process.
7. Quality is as important as quantity. This applies to the foods the patient puts into her/his body (nutrient-rich or nutrient-empty), the kind of air s/he breathes (clean or polluted), the environment in which s/he lives (toxin-free or toxic), and the kind of relationships in her/his life (supportive or destructive), etc.
8. Naturopathic medicine recognizes and seeks to enhance three kinds of effects. *Specific* effects result from the application of a therapy that is known to have a particular effect on a targeted area (e.g., ingestion of a botanical substance that is known to reduce gastric acidity). *General* effects result from individualized treatment prescriptions, and serve to increase the body's resistance to external disruptions (e.g., the use of Astragalus to stimulate the immune system). *Non-specific* effects are those effects that are a result of the mind/body connection.

Characteristics and Data

In 2000, the University of California, San Francisco Center for the Health Professions reported that there were approximately 2,000 licensed naturopathic physicians practicing in North America[4]. Since then, the profession has more than doubled. In 2016 the American Association of Naturopathic Physicians (AANP) estimated that there were approximately 5,000 naturopathic physicians licensed/registered in the United States and its territories, and the Canadian Association of Naturopathic Doctors stated that there were 2,357 naturopathic doctors registered in the Canadian provinces and territories. Of the respondents in the most recent (2015) Association of Accredited Naturopathic Medical Colleges (AANMC) Alumni Survey, 76% were female, the mean age was 41 (median 39). From 2000 to 2015, the median age of graduation was 30 or under. Seventy-two percent graduated in four years or less and 86% passed NPLEX Part 2 board examinations on the first attempt. Of the survey respondents, approximately 84% were practicing in regulated states/provinces.[5]

According to the *Chronicle of Higher Education's* 2009 Occupational Brief, naturopathic physicians work 30 to 50 hours a week.[6] Of the respondents in the 2015 AANMC Alumni survey, 92% said they were using their degree professionally. 49% were practicing full-time and 22% were practicing part-time. Of those practicing part-time, 34% were caring for family members or children. Most naturopathic physicians provide health care through office-based practice. Practitioners may make house calls or see patients in the evenings or on weekends. In addition to seeing patients, some naturopathic physicians serve as adjunct faculty at one of the naturopathic medical colleges or other health professions programs (e.g., nursing). Many lecture in other venues. Additionally, naturopathic doctors diversify their career paths as writers, health care entrepreneurs, contributors to media, business and wellness consultants, and in healthcare leadership roles.

Entry-level naturopathic physicians may earn $35,000 if starting in a part-time practice. Naturopathic residents average $35,000 for their first year. Naturopathic practices typically see income increase significantly with years in practice, with naturopathic physicians in established practices averaging $90,000. Net income is dependent on many factors, including the number of hours devoted to practice, the location of the practice, overhead expenses, and insurance coverage. Earning potential is also affected by professional certification, the types of services offered, the emphasis of treatment, and the population where the practice is centered.

Clinical Care

Approach to patient care

Using the six principles outlined above, naturopathic physicians seek to understand the factors that underlie and contribute to the patient's condition, to remove obstacles to healing, to strengthen the patient's inherent healing ability, and to teach the patient about dietary and lifestyle choices that support wellness and optimal health.[7] In striving to attain an in-depth understanding of the patient's health, naturopathic physicians consider genetic predispositions in combination with superimposed factors such as nutritional status, work and emotional stressors, environmental allergens and toxins, and biomechanical issues. Essential to a comprehensive evaluation is the extended interview, which ranges from 60 to 90 minutes for a new patient. A standard review of systems is supplemented with patient-generated reports of daily activities (including dietary habits, physical activity, and psychological well-being). Naturopathic physicians perform physical examinations and use conventional as well as innovative laboratory and diagnostic imaging procedures when appropriate. With many treatment options at their disposal, naturopathic physicians follow a distinct clinical rationale to individualize patient care within the framework of naturopathic principles. Follow-up appointments typically range from 30 to 45 minutes, and may occur more frequently in the beginning in order to support dietary and lifestyle changes.

In 1996 the AANP recommended that further work on practice principles move from the profession to the academic community. Clinical faculty and practitioners built on the core foundation throughout the 1990s. Three principles were identified that underlie the *clinical practice* of naturopathic medicine: [8,9]

1. Disease is characterized as a process rather than a pathologic entity;
2. focus is on the determinants of health rather than on pathology; and
3. a therapeutic hierarchy guides the course of treatment.

As taught in naturopathic medical schools, the therapeutic hierarchy is a guideline for applying various modalities and approaches according to the unique needs of the individual patient and the natural order of the healing process. The therapeutic hierarchy proposes the following order from most gentle to most invasive:

1. Establish the conditions for health by removing obstacles to healing.
2. Stimulate the self-healing mechanisms.
3. Support weakened or damaged systems or organs.
4. Address structural integrity.
5. Address pathology using specific natural substances, modalities, or interventions.
6. Address pathology using specific pharmacologic or synthetic substances.
7. Employ surgical correction and other invasive therapies that may have significant side effects.

For example, for a child who has recurrent otitis media, conventional treatment would consist of repeated courses of antibiotics and, in some cases, tympanostomy. A naturopathic physician would take a different approach:

1. Use the child's medical history, physical examination, and laboratory tests to look for and remove or address obstacles to health. In this child's case, these obstacles may include allergens (e.g., dust mites, dander, etc.), environmental irritants (e.g., second-hand cigarette smoke), food sensitivities (dairy, wheat, and eggs are most often implicated), poor nutrition (e.g., a diet high in processed grains and sugars), mechanical misalignments (e.g., of the cervical spine or cranial bones), and emotional factors (e.g., family stress). Children routinely improve significantly after these factors are addressed.
2. Stimulate the healing power of nature with therapies such as homeopathy and hydrotherapy.

3. Strengthen the affected systems by providing general immune support with nutritional supplementation (e.g., vitamins C and A) or botanical medicines (e.g., Echinacea angustifolia).
4. Address structural factors (e.g., using soft tissue manual therapy such as lymphatic drainage).
5. Treat the pathology with specific natural therapies (e.g., topical garlic, *Verbascum Thapsus* oil, *Hypericum perforatum* oil).
6. Prescribe a course of antibiotics if the infection is particularly recalcitrant or severe, or if there are other compromising factors.
7. Refer the child to a pediatrician or otolaryngologist if the infection still does not subside.

The therapeutic hierarchy creates a guideline for treatment that is both consistent with the principles of naturopathic medicine and addresses the patient's dynamic needs.[10]

Additionally, the Council of Chief Academic and Clinical Officers (CCACO) of the AANMC completed the AANMC Professional Competencies of the Graduating Naturopathic Physician in 2014.[11] The purpose of the document is to describe the core competencies of a graduate from an accredited naturopathic doctoral program in order to align curriculum, define expectations of graduates and inform stakeholders regarding the education of physicians who practice naturopathic medicine. The expectation is that this document will serve to guide current and future programs of naturopathic medical education.

Scope of practice

Naturopathic physicians are trained as primary care providers (PCPs). Some choose to specialize in populations, modalities, or other focused areas of clinical practice. Every jurisdiction that licenses/regulates the practice of naturopathic medicine gives naturopathic physicians the authority to diagnose as well as treat. This ability to diagnose distinguishes naturopathic physicians from practitioners who use natural therapies (such as botanical medicines and homeopathy) but who were not trained in naturopathic medical programs accredited by the Council on Naturopathic Medical Education (CNME), and who are not licensable in any of the regulated jurisdictions. Diagnosis includes eliciting a medical history, performing a physical examination, and (in many cases) ordering laboratory tests and/or diagnostic imaging studies.

In the nineteen states and territories that license or otherwise regulate the practice of naturopathic medicine, NDs are empowered to prevent, diagnose, and treat patients' health conditions. This includes the ability to perform physical examinations and order labs and diagnostic imaging. NDs are allowed to prescribe botanical medicines, nutritional supplements, and homeopathic remedies, to utilize hydrotherapy and physical therapies, and to provide dietary and lifestyle counseling. In three states (Oregon, Vermont, and Washington), NDs practice as full scope primary care providers, with several other states (Arizona, California, Connecticut, Montana, New Hampshire, Utah) coming close. Nine states allow naturopathic physicians to practice minor surgery, and ten states have drug prescription rights that allow some type of parenteral (IM or IV) administration of supplements or drugs. In some states, prescriptive authority is limited to first-line antibiotics and bio-identical hormones, while in other states NDs have full prescriptive authority for all drugs except chemotherapeutics and anti-psychotics. Prescriptive authority is important for continuity of care, as a naturopathic physician cannot take a patient off a drug unless s/he also has the right to prescribe that drug. Several states allow for additional scope of practice in specific areas under the naturopathic license. For example, six states allow naturopathic doctors to practice acupuncture, and six states plus the District of Columbia allow them to attend childbirth

Referral practices

Naturopathic physicians are trained and may serve as primary care providers. They collaborate with other health professionals and refer patients for optimum management of the patient's health care, especially for those patients who have life-threatening or challenging chronic conditions.[12] Naturopathic physicians are also trained to work within their scope of practice and refer, as appropriate, when patient demands fall outside of their skill or ability to treat.

Third-party payers

Insurance reimbursement for naturopathic physicians in the United States varies from jurisdiction to jurisdiction, although in the regulated states many private and governmental programs cover naturopathic care. The NPLEX PA survey found that only 19% of patient visits in North America were covered by insurance, although it must be kept in mind that not all naturopathic physicians accept third-party insurance. Some states provide student loan forgiveness to naturopathic physicians for service to urban and rural underserved communities and include naturopathic physicians as primary care providers in medical home legislation. In Vermont, naturopathic care is covered under Medicaid. In Oregon, naturopathic physicians are included in some coordinated care organizations. In Canada, naturopathic services are not covered under the publically funded health care system with the exception of British Columbia, which provides very limited coverage for those on premium assistance or social assistance. Coverage is provided by virtually all private insurance companies either under Extended Health Care Plans or Health Spending Accounts. Under the former coverage is generally $300 to $500 per year. Health Spending Accounts provide 100% coverage for services provided by Naturopathic Doctors. The number of Canadians with Extended Health Care coverage and/or Health Spending Accounts is on the increase.

Integration Activities

Naturopathic physicians practice in a wide variety of clinical settings and collaborate with diverse practitioners from both conventional and integrative health professions. The AANMC Alumni survey found that the majority of participants own or co-own a practice or business (67%). Of those, 45% own a solo private practice out of a clinic or office. 21% own a private multidisciplinary group practice, and 14% own a private group practice with other NDs. Those who are employees are most likely to work in a private multidisciplinary group practice (39%), or a private group practice with other NDs (30%). A significantly higher percentage of those working in licensed areas work in private group practices with other NDs (37% vs. 16%). A significantly higher percentage of those working in areas not requiring a license work in hospitals (32% vs. 4%).

Examples of integration with more conventional medical practitioners include:

- Integrative oncology: A number of oncology centers around the country integrate naturopathic physicians into their hospital and outpatient centers, providing patients with a wide range of treatment options and expertise in combating their illness.
- Integrative rheumatology: Many patients suffering from autoimmune diseases such as rheumatoid arthritis and systemic lupus erythematosus benefit by combining the skills and knowledge of a conventional rheumatologist (MD) and a naturopathic physician.
- Two Institute of Medicine projects in 2009-2011 included naturopathic physicians either on committees or contracted with them for papers.
- Public, rural, and community health: public and community health clinics in licensed states with the broadest scope of practice often integrate services from naturopathic physicians or utilize NDs as primary care physicians.
- For international communities in need (e.g., Nicaragua, Haiti, Mexico, Africa, India), healthcare services are being provided through naturopathic college-based student preceptorships and internships offered by organizations such as Naturopathic Doctors International and Naturopathic Medicine for Global Health.

Education

Schools and Programs

As of fall 2016, approximately 2257 students are currently enrolled at one of seven schools that offer CNME-accredited naturopathic medical programs:

- Bastyr University in Seattle, Washington, founded in 1978

  - Including Bastyr University branch campus in San Diego, California, opened in 2012
- Boucher Institute of Naturopathic Medicine in Vancouver, British Columbia, founded in 1998
- Canadian College of Naturopathic Medicine in Toronto, Ontario, founded in 1978
- National University of Natural Medicine in Portland, Oregon, founded in 1956 (formerly National College of Naturopathic Medicine)
- National University of Health Sciences in Lombard, Illinois, Naturopathic Program started in 2004
- Southwest College of Naturopathic Medicine and Health Sciences in Tempe, Arizona, founded in 1992
- University of Bridgeport College of Naturopathic Medicine in Bridgeport, Connecticut, Program started in 1997

There is one school that is a candidate for accreditation:

- Universidad del Turabo in Gurabo, Puerto Rico, Naturopathic Program started in 2008

Student Population

For students enrolled in the Fall 2015 entering class, the median cumulative GPA was 3.335. The mean age was 28, with a bi-modal distribution of students in their early 20s continuing directly after their undergraduate programs, and older students pursuing a second or third career. Minority students comprised approximately 22% of the 2011 entering class. The typical naturopathic student population is approximately 76% female.

Faculty

Faculty in the biomedical sciences predominantly have PhDs. Clinical faculty at naturopathic medical schools must have a terminal degree in their respective fields. The clinical sciences are taught by naturopathic doctors and other licensed practitioners such as medical doctors, osteopathic doctors, chiropractors, psychologists, and practitioners of acupuncture and Oriental medicine.

Curriculum content

Naturopathic physicians are trained through an educational process that is comparable to that of allopathic medical doctors (MDs) and doctors of osteopathy (DOs). Acceptance into a naturopathic medical school requires a bachelor's degree with pre-medical sciences, including biology, general and organic chemistry, and physics, as well as psychology and humanities. The naturopathic college admission process requires official undergraduate transcripts and essays that provide insight into the candidate's understanding of naturopathic medicine, as well as her/his motivation for applying to naturopathic medical school. Individuals who meet the application criteria are interviewed by faculty, staff, and students to assess academic ability, interpersonal skills, and professional demeanor.

With the biomedical sciences as a foundation, naturopathic medical education integrates Western diagnostic decision-making skills with natural as well as conventional therapies. Students spend more than 4,000 hours during the four-year post baccalaureate program learning the art and science of naturopathic medicine. Naturopathic principles, philosophy, and theory guide the curriculum and provide a conceptual framework for students to develop a profound understanding of humans in health and disease. The curriculum is dynamic, adapting to meet the changing healthcare needs of the population and to address the many obstacles to healing that patients experience. Naturopathic medical education pays particular attention to the growing epidemic of chronic diseases that affect every age group and sector of society.

At naturopathic medical school, students take two years of graduate level studies in the biomedical sciences. Although naturopathic medical students receive training in the same biomedical sciences in which allopathic medical students are trained, there are differences in the emphasis in the courses and in the way that knowledge is applied. For example, slightly less emphasis is placed on anatomy, requiring an average of 350 hours in a naturopathic curriculum, compared to 380 hours in an allopathic/osteopathic curriculum. However, in naturopathic programs, greater emphasis is placed on physiology. Naturopathic students receive approximately twice as many hours of physiology as allopathic students (250 hours compared to 125 hours)[13]. This difference may reflect the greater emphasis in naturopathic medicine on working with the dynamic processes of the person, rather than on the static structure of the body.

Coursework during the first year focuses on developing students' understanding of the human body in health and disease. Recent trends show programs moving toward integrated systems-based biomedical training, with clinical education introduced in the first year or two. Anatomy, biochemistry, microbiology, physiology, embryology, histology, neuroanatomy, and genetics accompany introductory classes in naturopathic philosophy, nutrition, mind-body medicine, homeopathy, and botanical medicine. The second year introduces students to Western diagnostic knowledge, with courses in clinical diagnosis, pathology, lab diagnosis, and diagnostic imaging. Lab sections teach students critical skills in history-taking and physical examination. Intermediate courses in naturopathic therapeutics continue to deepen students' understanding of clinical nutrition, manipulative therapy and physical medicine, homeopathy, and botanical medicine. In years three and four, students engage in supervised patient care on diverse patient populations. Didactic education continues to build expertise in naturopathic therapeutics, and adds in-depth coursework in pediatrics, gynecology, gastroenterology orthopedics, cardiovascular health, disorders of the eyes, ears, nose, and throat, oncology, nephrology, and dermatology. Naturopathic principles guide the curriculum design and course content.

Clinical training takes place in a variety of clinical settings. Clinical faculty provide training and supervision of care in multidisciplinary medical centers. Students see patients who have a wide range of clinical conditions, from acute illnesses (e.g., respiratory infections, influenza, gastrointestinal infections, musculoskeletal injuries, and minor lacerations) to chronic, sometimes life-threatening diseases (e.g., asthma, diabetes, colitis, heart disease, hypertension, hyperlipidemia, metabolic syndrome, arthritis, cancer, kidney disease, etc.). Experienced licensed clinicians, including naturopathic doctors, medical doctors, osteopaths, chiropractors, psychologists, and practitioners of acupuncture and Oriental medicine supervise students and oversee high quality patient care in the school clinics. In addition to the colleges' outpatient clinics and medical centers, students provide free care to thousands of medically underserved women, children, and men at homeless shelters, HIV/AIDS clinics, drug and alcohol rehabilitation programs, elementary schools in impoverished immigrant communities, shelters for battered women and children, and community health centers. These programs benefit the patients, who receive high quality naturopathic, conventional, and preventive care, as well as the students, who are introduced to a diverse population of patients who are often under-insured and suffering disproportionately from both chronic and acute conditions.

Accreditation

The Council on Naturopathic Medical Education (CNME) was recognized by the US Department of Education in 2003 as a programmatic accrediting agency, and has been recognized continuously since then. CNME's scope of recognition is the accreditation and preaccreditation throughout the United States and Canada of graduate-level, four-year naturopathic medical education programs leading to the Doctor of Naturopathic Medicine (NMD) or Doctor of Naturopathy (ND). In 2015, the US Department of Education re-recognized the CNME for a period of five years, the maximum time allowed.

The CNME Board comprises institutional members from accredited naturopathic medical schools, profession members who are licensed naturopathic physicians and public members who possess educational experience and are unaffiliated with any naturopathic program or the profession. In addition to establishing educational standards for accreditation, the CNME conducts a rigorous accreditation process similar to other medical fields; this process includes completion of a detailed self-study report that demonstrates compliance with the CNME accreditation standards, as well as an onsite visit by a team of experts to verify compliance first-hand. Onsite visits are conducted before granting candidacy status and initial accreditation to a naturopathic medical program, and are also conducted subsequently at least once every seven years as the basis for renewing accreditation status.[14]

In addition to having programmatic accreditation by the CNME, all colleges and universities in the United States that offer naturopathic medical programs have institutional accreditation from their respective regional accrediting agencies; in Canada, the institutions offering these programs are approved by their respective provincial higher education regulatory bodies. Accreditation holds institutions to high academic, financial, and human resource standards. Accreditation also ensures that schools respect academic freedom in teaching and include faculty, staff, and students in decision-making processes. Equally important, accredited institutions join a community of colleges and universities who share the ideas, innovations, and experiences that drive continuous quality improvement.

As noted above, seven naturopathic programs are currently accredited by the CNME, with one additional school in candidacy status.

Regulation and Certification

Regulatory status

The Federation of Naturopathic Medicine Regulatory Authorities was established in 2011 to facilitate communication between regulatory bodies, and to serve as a single body with which other entities in the profession can communicate. In the United States, naturopathic medicine is regulated in 19 states (Alaska, Arizona, California, Colorado, Connecticut, Hawaii, Kansas, Maine, Maryland, Massachusetts, Minnesota, Montana, New Hampshire, North Dakota, Oregon, Pennsylvania (law takes effect 2018), Utah, Vermont, and Washington) and the District of Columbia. Two territories, Puerto Rico and the Virgin Islands, are also regulated. Licensure efforts are underway in at least 12 other states. In Canada, naturopathic medicine is regulated in six provinces (Alberta, British Columbia, Manitoba, Nova Scotia, Ontario, and Saskatchewan). A high priority for both the American Association of Naturopathic Physicians and the Canadian Association of Naturopathic Doctors is to support legislative efforts to increase the number of states and provinces that regulate the profession.

To be eligible to be licensed/registered as a naturopathic physician in a regulated jurisdiction, a candidate must graduate from an accredited naturopathic medical school, and must pass national board examinations, other jurisdiction-specific examinations, and background checks.

Board examinations

The North American Board of Naturopathic Examiners (NABNE) is responsible for ensuring that candidates for licensure/registration have the requisite knowledge and skills to be safe practitioners. NABNE administers the Naturopathic Physicians Licensing Examinations (NPLEX), which are the board examinations required by all jurisdictions that regulate naturopathic medicine.

NPLEX is administered in two parts. Both parts are competency based. Students take the Part I Biomedical Science Examination after they complete their biomedical science training. Part II Clinical Sciences Examinations may be taken after the candidate has passed the Part I Examination and has graduated from a CNME-accredited naturopathic medical school.

NPLEX Part I—Biomedical Science Exam Areas

Anatomy
Biochemistry & Genetics
Microbiology & Immunology
Pathology
Physiology

NPLEX Part II— Clinical Science Examinations

Part II Core Clinical Science Exam Areas:

Physical & Clinical Diagnosis
Laboratory Diagnosis & Diagnostic Imaging
Botanical Medicine
Clinical Nutrition
Physical Medicine
Homeopathy
Health Psychology & Counseling
Research

Emergency Medicine & Medical Procedures
Pharmacology

Part II—Elective Examinations
Minor Surgery
Acupuncture

A new elective examination will first be administered in August 2017. The Pharmacology Elective Examination will focus on testing the knowledge NDs need if they are going to practice in jurisdictions that grant prescriptive (drug) authority to licensed NDs.

The clinical examinations are based on a practice analysis of the profession, the most recent of which was completed in 2011. Both the Part I and Part II Examinations use an integrated case-based format. For the Part I Biomedical Science Examination, a brief case is presented and the examinee is asked four or five questions regarding anatomy, physiology, biochemistry, pathology, microbiology and/or immunology that are relevant to the patient's diagnosis. For the Part II Core Clinical Science Examination, a more complete case is presented (including results of lab testing and diagnostic imaging), and the examinee is asked to diagnose the condition, interpret lab and imaging findings, and prescribe appropriate treatments within the context of naturopathic principles.

Research

Clinical trials and some laboratory studies are conducted at naturopathic medical schools in the United States and Canada. As the profession matures and the capacity for research expands (reflected in institutional commitment and development of a skilled research workforce), the number and quality of research studies improve. Naturopathic doctors often earn an MPH (Masters in Public Health) or PhD degree as part of their development as researchers, and then work at naturopathic institutions or at conventional academic health centers. There are currently naturopathic doctor researchers at the University of Washington, Columbia University, Yale University, the

University of Michigan, and Oregon Health and Science University. Others work with agencies such as the Group Health Research Institute, National Institutes of Health, and the RAND Corporation.

From 2002-2004, the National Institutes of Health (NIH) National Center for Complementary and Alternative Medicine (NCCAM, now known as National Center for Complementary and Integrative Health, NCCIH) funded the development of the Naturopathic Medical Research Agenda (NMRA). The NMRA brought together naturopathic physicians and conventional research scientists to identify and prioritize a list of research questions and areas.[15] Research departments from all naturopathic medical schools participated in a series of meetings and produced a consensus document on the future of naturopathic research. The highest priorities focused on three areas:

- Naturopathic treatment of type 2 diabetes
- Naturopathic care for the preservation and promotion of optimal health in geriatric populations
- Developing methodologies to understand the healing process

Between 1999 and 2012, NCCAM (now NCCIH) awarded approximately $25 million in research grants to institutions with programs in naturopathic medicine. Studies have been published on naturopathic approaches to specific conditions (e.g., diabetes and cardiovascular disease), the effectiveness of individual botanical medicines (e.g., *Urtica urens* for hay fever, *Echinacea angustifolia* for URIs, *Cimecifuga racemosa* for menopausal symptoms); homeopathic medicines (e.g., coca for altitude sickness, regional pollens for hay fever); nutritional supplementation (e.g., SAMe for osteoarthritis) and naturopathic detoxification for environmental illnesses. Other published studies have examined the nature of patient experience with naturopathic treatment.

In 2010, the Naturopathic Physicians Research Institute (NPRI) was founded. A key focus in this organization has been "researching the way we practice" using research methods which recognize that most disease has multiple causes and is best addressed using an array of interventions. Virtually all naturopathic practice reflects this perspective; consequently, single agent trials do not adequately measure the value of naturopathic medicine or support enhancement of its practice. The NPRI website lists some of these emerging whole practice trials. Naturopathic researchers were among the leaders who played a significant role in a successful ACIH-led effort to educate the NCCIH to elevate the importance of examining the impact of whole disciplines in the agency's 2011-2015 strategic plan.[16]

Challenges and Opportunities

Key challenges 2016-2020

- The greatest challenge facing the naturopathic profession today is the insufficient number of states and provinces that regulate naturopathic medicine. Currently less than a third of the states regulate the profession. With licensed naturopathic physicians practicing in only 19 regulated states, the need for what naturopathic medicine has to offer cannot be met nationally. Furthermore, healthcare consumers in the 31 unregulated states do not have adequate knowledge of naturopathic medicine. This has implications not only for patient demand, but for student recruitment as well. The profession must continue to push for regulation in more states and provinces.
- Consumers are confused regarding the differences between licensed naturopathic doctors and those practitioners (in unregulated states) who call themselves naturopaths or NDs but who do not have professional degrees or the credentials to be safe and effective practitioners. The profession must focus on educating consumers about the value of trained, licensed naturopathic physicians.
- Scope of practice in some jurisdictions is limited in comparison to the knowledge and skills taught in accredited naturopathic programs. This limitation impedes practitioner effectiveness in addressing patient issues in varying degrees.
- Although the number of accredited naturopathic medical programs has increased significantly, there are approximately

7,500 naturopathic physicians in practice today in North America. The small number of naturopathic practitioners clearly limits the opportunity to play a more significant role in the growing popular demand for the integrative medical practices in which naturopathic physicians are trained. This void is increasing being filled by other providers, including health coaches and Functional Medicine physicians.

Key opportunities 2016-2020

The capacity to train naturopathic physicians has increased significantly (the number of naturopathic medical schools has more than doubled in the past 20 years), and graduating class sizes are increasing. There is a larger and better qualified applicant pool available to the schools. Competencies and outcomes have been clearly defined for accredited schools and are regularly updated as the profession develops. All of these factors position the profession to take advantage of the opportunities that are presenting themselves, including:

- There is a growing need for primary care providers, particularly in light of the increased number of Americans who have access to health care due to the 2010 Affordable Care Act. The naturopathic profession's greatest strength lies in its ability to help address this need. Furthermore, the number of conventional medical students pursuing primary care/family care has been decreasing, leaving a void which NDs can help fill.
- There is an emerging epidemic of chronic diseases traceable to diet, environment, and lifestyle (e.g., diabetes, hypertension, heart disease, gastrointestinal disorders, environmental illnesses, allergies, etc.). Recognition by even the conventional healthcare profession of these etiologic factors is making it clear that patients must accept responsibility for their own health; it is vital that patients have physicians with whom they can partner on this journey. Naturopathic physicians are trained to work with their patients, spending time to educate them on lifestyle factors.

Consequently, the services naturopathic physicians have to offer are in greater demand than ever before.
- There are diverse practice opportunities for naturopathic physicians that were unheard of 25 years ago. Naturopathic physicians increasingly work in clinical settings with medical and osteopathic doctors, chiropractors, acupuncturists, massage therapists, and other IHM practitioners. Naturopathic physicians can be found in tribal clinics on Native American lands, in public health clinics, in cancer treatment centers, in integrative medical centers, in mental health clinics, in substance abuse programs, in corporate wellness programs, and at resorts and spas.
- Collaboration is increasing among schools and other professional ND organizations for the growth and advancement of the profession and for greater accessibility for patients.
- State and national healthcare reform are providing opportunities for greater integration of naturopathic physicians into the mainstream healthcare system.

Resources

Organizations and websites

National Associations:
- American Association of Naturopathic Physicians
  www.naturopathic.org
- Canadian Association of Naturopathic Doctors
  www.cand.ca (also www.naturopathicassoc.ca)

Education and Research:
- Association of Accredited Naturopathic Medical Colleges
  www.aanmc.org
- Council on Naturopathic Medical Education
  www.cnme.org
- Bastyr University
  www.bastyr.edu

- Boucher Institute of Naturopathic Medicine
  www.binm.org
- Canadian College of Naturopathic Medicine
  www.ccnm.edu
- National University of Natural Medicine
  www.nunm.edu
- National University of the Health Sciences
  www.nuhs.edu
- Southwest College of Naturopathic Medicine
  www.scnm.edu
- University of Bridgeport College of Naturopathic Medicine
  www.bridgeport.edu/academics/graduate/naturo
- Naturopathic Post-Graduate Association
  www.np-ga.org
- Naturopathic Medical Student Association
  www.naturopathicstudent.org
- Naturopathic Physicians Research Institute
  www.nprinstitute.org

Regulation:
- Federation of Naturopathic Medicine Regulatory Authorities
  www.fnmra.org
- North American Board of Naturopathic Examiners
  www.nabne.org

Specialty Associations:
- American Association of Naturopathic Midwives
  www.naturopathicmidwives.com
- Homeopathic Academy of Naturopathic Physicians
  www.hanp.net
- Oncology Association of Naturopathic Physicians
  www.oncanp.org
- Pediatric Association of Naturopathic Physicians
  www.pedanp.org
- Institute of Naturopathic Generative Medicine
  www.generativemedicine.org/INGM

Other Organizations:

- Natural Doctors International
  www.ndimed.org
- Naturopathic Medicine for Global Health
  www.natmedglobalhealth.org

Bibliography

General Naturopathic Medicine:

Alschuler LN, Gazella KA. *Alternative Medicine Magazine's Definitive Guide to Cancer: An Integrative Approach to Prevention, Treatment, and Healing*. Berkeley, CA: Celestial Arts; 2007.

Bove M. *An Encyclopedia of Natural Healing for Children and Infants*. New York, NY: McGraw-Hill; 2001.

Canadian College of Naturopathic Medicine Press. Fifteen textbooks based on naturopathic curriculum, written by naturopathic physicians, faculty, and lecturers. http://www.ccnmpress.com/. Accessed November 17, 2009.

Hudson T. *Women's Encyclopedia of Natural Medicine: Alternative Therapies and Integrative Medicine*. Lincolnwood, IL: Keats; 1999.

Pizzorno JE, Murray MT, Joiner-Bey H. *The Clinician's Handbook of Natural Medicine*. 2nd ed. Philadelphia, PA: Churchill Livingston; 2002.

Pizzorno JE, Murray MT. *The Encyclopedia of Natural Medicine*. 2nd ed, revised. Roseville, CA: Prima Publishing; 1997.

Pizzorno JE, Murray MT. *The Textbook of Natural Medicine*. 4th ed. Philadelphia, PA: Elsevier; 2013.

Pizzorno JE. *Total Wellness: Improve Your Health by Understanding the Body's Healing Systems*. Roseville, CA: Prima Publishing; 1997.

Standish L, Calabrese C, Snider P. *The Future and Foundations of Naturopathic Medical Research Science: Naturopathic Medical Research Agenda*. Kenmore, WA: Bastyr University Press; 2005.

Yarnell E. *Naturopathic Gastroenterology*. East Wenatchee, WA: Healing Mountain Publishing Inc; 2011.

Naturopathic Philosophy:

Boyle W, Saine A. *Lectures in Naturopathic Hydrotherapy.* East Palestine, OH: Buckeye Naturopathic Press; 1988.

Kirchfeld F, Boyle W. *Nature Doctors: Pioneers in Naturopathic Medicine.* Portland, OR: Medicina Biologica; 1994.

Kneipp S. *My Water-Cure.* New ed. Champaign, IL: Standard Publications Inc; 2007.

Lindlahr H. *Nature Cure.* Charleston, SC: BiblioBazaar LLC; 2007.

Lindlahr H. *Philosophy of Natural Therapeutics.* Champaign, IL: Standard Publications Inc; 2007.

Lust B. *Collected Works of Dr. Benedict Lust.* East Wenatchee, WA: Healing Mountain Publishing Inc; 2006.

Smith F. *An Introduction to Principles and Practice of Naturopathic Medicine.* Toronto, Ontario, Canada: CCNM Press; 2008.

Zeff JL. The process of healing: a unifying theory of naturopathic medicine. *J Naturopathic Med.* 1997;7(1):122-5.

Zeff J , Snider P, Myers S, DeGrandpre Z. A hierarchy of healing: the therapeutic order. The unifying theory of naturopathic medicine. In: Pizzorno JE, Murray MT. *Textbook of Natural Medicine.* 4th ed. Philadelphia, PA: Elsevier;2013, 18-33.

Other:

Chaitow L, Blake E, Orrock P, Wallden M, Snider P, Zeff J. *Naturopathic Physical Medicine: Theory and Practice For Manual Therapists and Naturopaths.* Philadelphia, PA: Churchill Livingston Elsevier; 2008.

Dooley TR. *Homeopathy Beyond Flat Earth Medicine.* 2nd ed. San Diego, CA: Timing Publications; 2002.

Marz RB. *Medical Nutrition from Marz.* Portland, OR: Quiet Lion Press; 1997.

Mitchell W. *Plant Medicine in Practice: Using the Teachings of John Bastyr.* Philadelphia, PA: Churchill Livingston Publishing; 2003.

Murray M, Pizzorno J, Pizzorno L. *The Condensed Encyclopedia of Healing Foods.* New York, NY: Atria Books; 2006.

Neustadt J. *A Revolution in Health through Nutritional Biochemistry.* Bloomington, IN: iUniverse, Inc; 2007.

Stargrove M, Treasure J, McKee DL. *Herb, Nutrient, and Drug Interactions: Clinical Implications and Therapeutic Strategies.* St. Louis, MO: Mosby; 2007.

Tilgner S. *Herbal Medicine from the Heart of the Earth.* Pleasant Hill, OR: Wise Acres; 1999.

Journals and Periodicals:

International Journal of Naturopathic Medicine. http://www.intjnm.org. Peer reviewed online journal.

Natural Medicine Journal. www.naturalmedicinejournal.com/. Peer reviewed online journal.

Naturopathic Doctor News and Review. http://www.ndnr.com. Professional news and information resource for naturopathic physicians in North America.

Citations

1. House of Delegates Position Paper Definition of Naturopathic Medicine. American Association of Naturopathic Physicians website. http://www.naturopathic.org/files/Committees/HOD/Position%20Paper%20Docs/Definition%20Naturopathic%20Medicine.pdf
2. Hough H, Dower C, O'Nell E. *Profile of a Profession: Naturopathic Practice.* San Francisco, CA: University of California at San Francisco Center for the Health Professions; 2001.
3. American Association of Naturopathic Physicians. Select Committee on the Definition of Naturopathic Medicine, Snider P, Zeff J, co-chairs. Definition of Naturopathic Medicine. Position Paper. Rippling River, OR; 1989.
4. Albert D, Martinez D. The supply of naturopathic physicians in the United States and Canada continues to increase. *Comp Health Prac Rev.* 2006; 11(2):120-122.
5. Naturopathic Physicians Licensing Examinations (NPLEX) 2011 Practice Analysis of the Naturopathic Profession. Portland, OR. Mountain Measurement Inc. http://www.mountainmeasurement.com/contact.php

6. Chronicle Guidance Publications. Occupational Brief 624: Naturopathic Physicians. http://www.chronicleguidance.com/store.asp?pid=7330 Accessed November 17, 2009.
7. Dunne N, Benda W, Kim L, et al. Naturopathic medicine: what can patients expect? *J Fam Pract.* 2005;54(12).
8. Zeff J. The process of healing: a unifying theory of naturopathic medicine. *J Naturopathic Med.* 1977;7(1):122-125.
9. Zeff J , Snider P, Myers S, DeGrandpre Z. A hierarchy of healing: the therapeutic order. The unifying theory of naturopathic medicine. In: Pizzorno JE, Murray MT. *Textbook of Natural Medicine.* 4th ed. Philadelphia, PA: Elsevier; 2013, 18-33.
10. Findings and Recommendations Regarding the Prescribing and Furnishing Authority of a Naturopathic Doctor. California. Dept. of Consumer Affairs, Bureau of Naturopathic Medicine; January 2007. http://www.naturopathic.ca.gov/formspubs/formulary_report.pdf
11. AANMC Professional Competencies of the Graduating Naturopathic Physician. Association of Accredited Naturopathic Medical Colleges website. http://aanmc.org/research-and-data/aanmc-professional-competencies-graduating-naturopathic-physicians/
12. Health Professions Advisors Guide. Champaign, IL: National Association of Advisors for the Health Professions Inc; 2007.
13. Jensen, C B. (1997). Common paths in medical education: the training of allopaths, osteopaths, and naturopaths. *Alternative and Complementary Therapies*.1997;3(4):276-280.
14. US Department of Education Office of Postsecondary Education. http://www.ed.gov/admins/finaid/accred/index.html Accessed November 17, 2009.
15. Standish L, Calabrese C, Snider P. The naturopathic medical research agenda: the future and foundation of naturopathic medical science. *JACM*. 2006;12(3):341-345.
16. Menard M, Weeks J, Anderson B, et al. Consensus Recommendations to NCCIH from Research Faculty in a Transdisciplinary Academic Consortium for Complementary and Integrative Health and Medicine. *JACM*. Jul 2015, 21(7): 386-394.

Section II

Related Integrative Practice Fields

Advanced Nutrition

Ayurvedic Medicine

Holistic Nursing

Homeopathy

Integrative Health and Medicine

Integrative Medicine

Yoga Therapy

Advanced Nutrition

Third Edition (2017) Authors: Jeffrey Blumberg, PhD, FASN, FACN, CNS-S and Dana Reed, MS, CNS, CDN

Partner Organization: Board for Certification of Nutrition Specialists (BCNS)

About the Authors: Blumberg is President and Reed is former director of the Board for Certification of Nutrition Specialists.

Philosophy, Mission, and Goals

The use of diet to treat disease originated in ancient Greece. During the 19th century, vitamin deficiency diseases were identified and many vitamins and essential minerals were defined in the early part of the 20th century. These discoveries led to implementation of diet-related guidelines by the military and recommended allowances for macro- and micronutrient intakes. The dietetics profession developed as an outgrowth of this work, and is largely oriented toward institutional food service management and acute disease settings.[1]

In the late 20th century, as evidence began demonstrating a powerful influence of food and nutrients on health and the pathophysiology of chronic disease,[2] the profession of advanced nutrition emerged. Based on the principles of nutrition science, including biochemistry, cellular and systemic metabolism, and physiology, advanced nutrition has become both a distinct profession and a specialty of other healthcare professionals.

A consortium of professional nutrition organizations is currently helping move nutrition from the periphery to the core of healthcare practice and policy, by establishing an independent career path for advanced nutrition professionals. The consortium includes:

- Disseminating unbiased nutrition science through the American College of Nutrition® (ACN), a professional society established in 1959. ACN publishes the peer reviewed *Journal of the American College of Nutrition,* provides continuing medical education (CME)

training and an organizational home for advanced nutrition professionals, both clinicians and researchers;
- Developing the competency standards and certifying the advanced nutrition professionals – Certified Nutrition Specialists® – through the Board for Certification of Nutrition Specialists (BCNS), established in 1993;
- Advocating for freedom and licensure to practice nutrition and for insurance reimbursement through the Center for Nutrition Advocacy®; and
- Educating practitioners and the public about nutrition for health through the American Nutrition Association (ANA).

Clinical Care

Approach to patient care

Advanced nutrition professionals work with patients and clients to prevent many diseases and conditions, and manage a broad range of chronic health conditions, including endocrine, digestive, cardiovascular, immune, bone, cancer, renal, cognitive and neurological, hormonal, urogenital, respiratory and other disorders.

Advanced nutritional professionals strive to treat each person based on his/her biochemical and genomic individuality, and often work in concert with other healthcare practitioners in patient management.

Scope of Practice

The regulatory scope of practice for advanced nutrition professionals varies from state to state. In the majority of states, the scope of practice includes:

- **Nutrition assessment** of dietary, anthropomorphic, laboratory, nutrigenomic, physical and other data. In some states, it includes ordering of diagnostic tests.
- **Nutrition intervention** designing and implementing a plan of action, which may include recommendations for food, medical

foods, or dietary supplements to prevent or manage a nutrition-related medical condition.

- **Nutrition counseling** providing nutrition counseling, education and support to prevent or manage nutrition-related conditions.

Referral practices

The advanced nutrition professional may work in a private setting or as a team member in an integrated medical clinic, or in academia and research, government, community nutrition, media, industry or other settings.

Clients may be referred by another healthcare practitioner or be self-referred.

Third-party payers

Insurance reimbursement for nutrition services is not yet universal. Medical nutrition therapy (MNT) for a wide range of chronic conditions is covered by many insurance plans or by health savings accounts, for both individual and group counseling. 'MNT' is the federally recognized phrase used in licensing regulatory language. Many insurance plans have a preventive medicine benefit, which covers MNT for hypertension, obesity, and diabetes. In most states, when medical nutrition therapy is covered by insurance, a referral from a provider such as a physician is required.

Education

Schools and programs

In order to acquire the skills and knowledge needed to be an advanced nutrition professional, a master of science or doctoral degree is required. Advanced degree programs in human nutrition, distinct from dietetics, are growing and new ones are under development.

Curriculum content

The curricula of programs in human nutrition vary but are generally focused on basic core competencies that are distinct from those of dietetics programs. Emphasis is on clinical biochemistry and metabolism, research methodology, epidemiology, and dietary intervention for optimal health and disease-specific care, including medical nutrition therapy, nutritional assessment, laboratory analysis, and developmental nutrition. Most programs address the evidence-based use of foods and nutraceuticals for optimizing health and managing specific health conditions. For basic educational requirements for eligibility for the CNS, please see the BCNS web site at *www.nutritionspecialists.org.*

Faculty

Generally, faculty members possess advanced and terminal degrees in their professional and academic fields and demonstrate relevant experience.

Accreditation

The standard for advanced degree programs in human nutrition is regional accreditation.

Regulation and Certification

Regulatory status

Regulations governing the practice of nutrition vary by state, and include a mix of exclusive scope of practice, certification, and registration. Currently, 22 states have exclusive scope of practice licensure, 25 states have title-protection-only licensure or certification, nine have only registration, and four states have no pertinent regulations. In all states that have regulation, nutrition and dietetics are combined. The CNS credential itself or the Certifying Examination for Nutrition Specialists meet the regulation requirements in many states. The most common title is certified or licensed dietitian/ nutritionist.

In exclusive scope of practice licensure states, it is illegal for those who are not licensed to practice nutrition. Title-protection-only licensure or certification states do not restrict the practice of nutrition, but only the titles that a practitioner may use. In some states, only licensed or certified practitioners are eligible to participate as private insurance providers.

The Center for Nutrition Advocacy® is the preeminent state and federal policy advocate for nutrition practitioners and has achieved much success in state licensing laws and federal regulations. It has been working successfully to enable CNSs and other qualified nutrition professionals to practice to the level of their training. Recent accomplishments include:

- Since 2011, defeating 23 bills to make practice exclusive to RDs in 12 states. Four states that formerly only recognized RDs, now recognize other nutrition professionals.[3] US Bureau of Labor Statistics incorporated changes to the Occupational Handbook entries for Nutrition and Dietetics in 2014 that recognize the CNS as the credential for advanced nutrition professionals and the RD as the credential for dietitians.[4]
- Centers for Medicare & Medicaid Services issued a ruling in which Medicare recognizes qualified nutritionists as equivalent to RDs in hospital and long-term care settings.[5]

Examinations and certifications

The Certified Nutrition Specialist® (CNS®) certification is the highest standard for the advanced nutrition professionals. The CNS candidates must meet rigorous requirements including: Master of Science or doctoral degree in nutrition or a related biomedical field from a regionally accredited university; coursework in advanced nutrition science and therapeutics, life sciences, with a focus on biochemistry; 1,000 hours of supervised practice experience; passing score on the Certifying Examination for Nutrition Specialists; and recertification every 5 years.

The CNS credential is conferred by the Board for Certification of Nutrition Specialists (BCNS). The CNS program is accredited by the National Commission for Certifying Agencies (NCCA).

The CNS credential and its Certifying Examination for Nutrition Specialists℠ are the most frequent non-dietetic credential and examination recognized in state nutrition laws and regulations.

Challenges and Opportunities

Advanced nutrition practice is a rapidly developing profession distinct from the field of dietetics. A consortium of national professional nutrition organizations has established advanced nutrition as a recognized, science-based profession. The growth of this field is evidenced by increases in the number of universities developing nutrition programs, candidates applying for CNS certification, increased membership in nutrition professional societies, and success of legislative efforts at both state and federal levels.

In addition to less than universal insurance reimbursement, state dietetics and nutrition practice laws remain a challenge, with some practice restriction existing for advanced nutrition professionals in 17 mostly less-populated states. However significant shifts in attitudes and court rulings regarding regulation have accelerated the momentum toward dismantling those remaining artificial barriers.

Resources

Organizations and Websites

- American College of Nutrition® (ACN) www.americancollegeofnutrition.org
- American Nutrition Association (ANA) www.americannutritionassociation.org
- Board for Certification of Nutrition Specialists (BCNS) www.nutritionspecialists.org
- Center for Nutrition Advocacy®, www.nutritionadvocacy.org

Citations

1. Whitney S, Rolfes S (ed): *Understanding Nutrition* 14th ed. Stamford, CT: Cengage Learning; 2016.
2. Murray C, Atkinson C, Bhalla K, et al. The State of US Health, 1990-2010: Burden of Diseases, Injuries, and Risk Factors, US

Burden of Disease Collaborators, JAMA. 2013;310(6):591-606. doi:10.1001/jama.2013.13805

3. The Center for Nutrition Advocacy State Regulatory Regime Analysis, December 2016
4. Bureau of Labor Statistics Occupational Outlook Handbook http://www.bls.gov/ooh/healthcare/dietitians-and-nutritionists.htm#tab-4
5. Centers for Medicare & Medicaid Services 42 CFR Parts 413, 416, 440, 442, 482, 483, 485, 486, 488, 491, and 493; Medicare and Medicaid Programs; Regulatory Provisions to Promote Program Efficiency, Transparency, and Burden Reduction; Part II, Final Ruling May 2014 p27118

Ayurvedic Medicine

First Edition (2009) Author, Second (2013) and Third (2017) Edition
Editor: Felicia Marie Tomasko, RN

Partner Organization: National Ayurvedic Medical Association (NAMA)

About the Author/Editor: Tomasko is an Ayurvedic professional and a member of the Board of Directors of the National Ayurvedic Medical Association, a member of the Board of Directors of the California Association of Ayurvedic Medicine and the Editor-in-Chief of *LA YOGA Ayurveda and Health* magazine.

Philosophy, Mission, and Goals

While Ayurveda as a system is in its relative infancy in the US as far as education, regulation, recognition, scope, and numbers of people practicing, it has a long and well-established historical tradition in its region of origin, the Indian subcontinent. A translation of the Sanskrit word Ayurveda is "science of life." The written source texts of this science are variously dated but are generally agreed to date to at least the beginning of the Common Era. Many scholars suggest the written tradition is older, or at least a compilation of a more ancient oral tradition. The three primary texts are the *Charaka Samhita,* the *Sushruta Samhita,* and *Vagbhata's Ashtanga Hridaya Samhita.* Information included therein is considered to be the genesis of classical Ayurveda. Other texts have added to the body of knowledge over time.

Ayurveda focuses on an individual's relationship with his/her own body, mind, and spirit, and with the natural world. These relationships begin with the five elements (earth, water, fire, air, and ether/space) and how they combine to create the three *doshas*: *vata* (air and ether/space), *pitta* (fire and water), and *kapha* (water and earth). Vata is light, dry, expansive, and changeable. Pitta is hot, intense, transformational, and sharp. Kapha is stable, heavy, solid, cold, and oily. Many physiological processes fall under these categories; movement is governed by vata, digestion and transformation by pitta, and insulation and lubrication by kapha.

According to Ayurveda, people have individual constitutional make-ups (*prakruti*), with unique proportionality of the doshas. When in balance, good health is experienced. When out of balance, disease is more likely to take hold. Importance is placed on disease prevention and maintaining balance with the ever-changing forces of nature. Balance is not static, but is constantly shifting; individuals must adjust their practices, routines, and lifestyles to maintain balance. Ayurveda is not exclusive of modern medicine, and combines well philosophically and practically because of its inclusivity.

Characteristics and Data

The National Ayurvedic Medical Association (NAMA) is currently collecting information on people practicing Ayurveda in the US and this data has not yet been publicly released. As of the 2015 NAMA Annual Report, the organization has 1089 members with 723 professional members. This represents a subsection of the individuals practicing Ayurveda in the US.

Scope of Practice

Defining scope of practice related to the modification of the traditional textual tradition to comply with modern integrative medical practice is a current project of NAMA. Scope of practice definitions for different levels of practitioner (Ayurvedic Health Counselor, Ayurvedic Practitioner and Ayurvedic Doctor) have been agreed upon by the NAMA Standards Committee and the community and are available on the NAMA website (www.ayurvedanama.org/page/ScopeforAyurveda). Scope of practice for associated practitioners, such as Pancha Karma Therapist or Technician and Ayurvedic Yoga Therapist, are under discussion by the NAMA Standards Committee.

Integration Activities

Many Ayurvedic professionals are also licensed healthcare providers in fields such as medicine, midwifery, chiropractic, nursing, massage therapy, and physical therapy. Ayurvedic professionals are often also yoga teachers and/or yoga therapists. Concepts of Ayurveda are taught in yoga teacher training programs and are part of the core curriculum in the standards of the International Association of Yoga Therapists as well as within the curriculum guidelines for Yoga Teachers approved by Yoga Alliance. Diverse collaborations are occurring in research settings, clinics, and hospitals.

Education

Ayurveda became initially visible in the US largely through efforts of spiritual communities. For example, in the 1970s, the Transcendental Meditation (TM) community brought doctors trained in India to the US to offer Ayurvedic consultations to meditators. Other notable yoga centers, such as the Himalayan Institute, Kripalu, and Mount Madonna Center, have brought Ayurvedic doctors from India to train students in the modality beginning in the 1970s and continuing through today. Esteemed doctors, including Indian natives who came to the US to practice, Westerners who studied in India, and practitioners who trained in the West, have furthered the profession in the US.

People practicing Ayurveda have varying educational backgrounds, including BAMS (bachelor of Ayurvedic medical science) degrees, MASc (masters in Ayurvedic science) degrees or, more recently, MD Ayurved (medical doctor of Ayurveda) degrees. A growing number of training programs exist outside the Indian system of education including certificate courses and master's level and clinical doctorate programs. Settings for these programs include independent Ayurvedic programs or institutions as well as programs taught at larger retreat and educational centers as well as at accredited four-year and graduate universities and chiropractic and other colleges. Each type enrolls and graduates students in the US. In addition to Western-style academic programs, all schools feature an internship component; some schools also have traditional apprenticeship models in addition to academic studies (above and beyond in-person internship required at all schools). Currently, schools are growing and expanding their internship programs. Representatives from 30 US-based schools have met to discuss the development of educational standards and scope of practice. As of 2016, 28 US-based schools are recognized as providing training at the professional membership level of Ayurvedic Health Counselor with NAMA. 20 US-based schools are recognized as providing training at the professional membership level of Ayurvedic Practitioner with NAMA. One US-based school is recognized as providing training at the professional membership level of Ayurvedic Doctor, and 13 US-based programs offer peripheral training. These schools meet current requirements, participate in the discussions on standards, and are eligible to participate in the annual NAMA meeting of schools.

Regulation and Certification

While Ayurveda is a licensed medical profession in its native India, and in countries including Nepal and Sri Lanka, Ayurveda is currently not licensed in the US. Yet self-regulation of the profession is increasing as organizations, such as NAMA, work on standards, scope of practice, recognizing practitioners, and developing a national exam. Several state organizations (in California, Colorado, Florida, Minnesota and Washington) have missions and goals involving licensure.

Research

A growing number of studies confirm the efficacy of many traditional treatments, such as the use of turmeric (*Curcuma longa*), a spice valued in Ayurveda for its anti-inflammatory and blood sugar regulating effects and bitter gourd (*Momordica charantia*) for diabetes prevention and treatment.

The Transcendental Meditation community sponsors hundreds of scientific studies investigating the use of meditation. Research into yoga therapy, which overlaps with Ayurveda, is on the rise. The study of Ayurvedic therapies using a Western approach is still developing when compared to other systems, and faces challenges as therapies are generally tailored to the individual rather than standardized for a particular condition.

Through the NAMA annual conference, the organization is actively supporting the community's participation in research through curation of poster presentations by longtime NAMA board member Jennifer Rioux, PhD (research focus includes weight loss as well as professionalism of Ayurveda, and development of research protocols), practicum sessions that focus on research methodology, and presentations of research. The Ayurveda Journal of Health focuses on Ayurvedic science, philosophy, and clinical practice. It is published by UMass Dartmouth. Diana Lurie, PhD, and Editor-in-Chief, is a Professor of Neuropharmacology at The University of Montana. Her research lab is currently investigating common herbs used in Ayurveda: *Bacopa monnieri*, Gotu Kola, and Ashwagandha, and their ability to affect the biochemistry of inflammatory processes. The Journal's editorial staff member Rammohan Rao, PhD, is a Research Associate Professor at the Buck Institute whose research on understanding the mechanisms of age and neurodegenerative diseases includes Ayurveda. Other members of the organization are continuing to promote an increase in research into Ayurveda treatments and systems methodology.

Challenges and Opportunities

Key challenges 2016-2020

Greater understanding and recognition of Ayurveda by consumers and the larger medical community are current challenges. Others include:

- identification of Ayurvedic practitioners at all levels currently active in the US and encouraging those who have not yet applied for professional recognition to do so
- continued unification of a dispersed and diverse community
- continued dissemination of the clear educational standards developed over the past few years that take into account differing training methodologies and lineages within Ayurveda
- continued dissemination of scope of practice guidelines for three different levels of practitioner with different amounts of training and ability to work with individuals and groups and are relevant in modern integrated healthcare practice
- strengthening relationships with other healthcare providers and organizations in integrated settings
- development and dissemination of research priorities and methodologies so that more of the Ayurvedic community is engaged in research
- communication among the Ayurvedic communities worldwide to develop common goals in furthering the profession
- continued coordination within the herbal community to support sustainability in obtaining farmed or gathered herbs as well as supporting appropriately labeled and safe imported herbal supplies
- beta testing and roll out of a standardized national examination program to support stronger professionalism
- support of greater professionalism of professional level members for the science of Ayurveda within the context of the modern healthcare climate

Key opportunities 2016-2020

Practices supporting the achievement of balance and optimal health can contribute to making a positive difference in the modern healthcare system. Opportunities include:

- teaching tangible, low-cost, self-implemented techniques for health maintenance and possible prevention of chronic diseases
- increased collaboration with other healthcare providers and professions

- combining the philosophy and therapies of Ayurveda with other systems to address often recalcitrant chronic illnesses
- continued support of healthcare freedom legislation as well as pursuing licensure
- increased membership in NAMA as well as increased attendance at conferences and other events as a means of networking and uniting the profession
- increased collaboration with related communities including Yoga Therapy, the teaching of yoga, massage therapy, nursing and medicine, chiropractic health care, allied sciences
- increased encouragement of and advocacy for research demonstrating the efficacy of Ayurveda as a system as well as Ayurvedic herbs, therapies, and treatments
- continued community involvement for individuals applying to be recognized at different levels of professional achievement (including Ayurvedic Educator, Practitioner, and Doctor)
- roll-out of national certifying exam and community involvement in testing process

Resources

Organizations and websites

The National Ayurvedic Medical Association is a 501(c)(6) organization and the largest Ayurvedic professional organization in the US. Membership as of 2015 is more than 100 members. NAMA holds an annual conference. NAMA has a council of schools that focuses on standards and scope issues. While NAMA may be the most visible organization of Ayurvedic practitioners, it does not necessarily represent everyone within the US Ayurvedic community. Website: www.ayurveda-nama.org. Some states have active organizations, notably:

- California Association of Ayurvedic Medicine (CAAM) www.ayurveda-caam.org/
- Washington Ayurvedic Medical Association (WAMA) www.ayurveda-wama.org/

- Colorado Ayurvedic Medical Association (Colorama)
  http://www.coloradoayurveda.org/
- Minnesota Ayurveda Association
  http://www.mnayurvedaassn.com

Holistic Nursing

Third Edition (2017) Editor: Carole Ann Drick, PhD, RN, AHN-BC

First Edition (2009) Author and Second Edition (2013) Editor:
Carla Mariano, EdD, RN, AHN-BC, FAAIM

Partner Organization: American Holistic Nurses Association (AHNA)

About the Author/Editors: Mariano developed and is former coordinator of the Adult Holistic Nurse Practitioner Program at New York University College of Nursing; developed the BS in Holistic Nursing Program at Pacific College of Oriental Medicine; and is past President of American Holistic Nurses Association. Drick is President of American Holistic Nurses Association.

Philosophy, Mission, and Goals

In 2006, holistic nursing was recognized by the American Nurses Association (ANA) as an official nursing specialty with a defined scope and standards of practice within the discipline of nursing. Holistic nursing emanates from five core values summarizing the ideals and principles of the specialty. These core values are:

- Holistic Philosophy, Theory, Ethics
- Holistic Caring Process
- Holistic Communication, Therapeutic Healing Environment, and Cultural Diversity
- Holistic Education and Research
- Holistic Nurse Self-Reflection and Self-Care

Holistic Nursing:
- embraces all nursing which has enhancement of healing the whole person across the lifespan and the health-illness continuum as its goal;
- recognizes and integrates body-mind-emotion-spirit, energetic-environment principles and modalities in daily life and clinical practice;

- focuses on protecting, promoting, and optimizing health and wellness, assisting healing, preventing illness and injury, alleviating suffering, and supporting people to find comfort, peace, harmony, and balance through the diagnosis and treatment of human response; and advocacy in caring for individuals, families, communities, populations, and the planet;
- views the nurse as an instrument of healing and a facilitator in the healing process;
- honors the individual's subjective experience about health, illness, health beliefs, and values;
- uses the caring-healing relationship and therapeutic partnership with individuals, families, and communities;
- draws on nursing knowledge, theories, research, expertise, intuition, and creativity;
- incorporates the roles of clinician, educator, consultant, coach, partner, role model, and advocate;
- encourages peer review of professional practice in various clinical settings and utilizes knowledge of current professional standards, laws, and regulations governing nursing practice;
- collaborates and partners with all constituencies in the health process including the person receiving care, family, significant others, community, peers, and other disciplines, using principles and skills of cooperation, alliance, consensus, and respect, and honoring the contributions of all;
- focuses on integrating self-reflection, self-care, and self-responsibility in personal/professional life;
- emphasizes awareness of the interconnectedness of self, others, nature, and God/Life/Spirit/Universal Force.

Characteristics and Data

There are 3.2 million nurses in the US, of which 7,000–12,000 identify as holistic nurses and 5,000 are members of the American Holistic Nurses Association (AHNA).

Scope of Practice

- draw on nursing knowledge, theories of wholeness, research and evidence-based practice, expertise, caring, and intuition to become therapeutic partners with clients and significant others in a mutually evolving process toward healing, balance, and wholeness;
- integrate holistic, complementary and integrative modalities including, for example, relaxation, meditation, guided imagery, breath work, biofeedback, aroma and music therapies, touch therapies, acupressure, herbal remedies and natural supplements, homeopathy, reflexology, Reiki, journaling, exercise, stress management, nutrition, self-care processes, and prayer with conventional nursing interventions;
- conduct holistic assessments of physical, functional, psychosocial, emotional, mental, sexual, cultural, age-related, spiritual, beliefs/values/preferences, family issues, lifestyle patterns, environmental, and energy field status;
- select appropriate interventions in the context of the client's total needs and evaluate care in partnership with the client;
- assist clients to explore self-awareness, spirituality, growth, and personal transformation in healing;
- work to alleviate clients' signs and symptoms while empowering clients to access their own natural healing capacities;
- concentrate on the underlying meanings of symptoms and changes in the client's life patterns;
- provide comprehensive health counseling, education and coaching, health promotion, disease prevention, and risk reduction;

- guide clients/families between the conventional allopathic medical system and complementary/integrative therapies and systems;
- collaborate with and refer to other healthcare providers/resources as necessary;
- advocate to provide access to and equitable distribution of healthcare resources, and to transform the healthcare system to a more caring culture;
- participate in building an ecosystem that sustains the well-being of the environment and the health of people, communities, and the planet;
- practice in numerous settings including acute, ambulatory, community, and home care, private practice, wellness/ complementary/integrative care centers, women's health centers, schools, employee/student health, psychiatric mental health facilities, rehabilitation centers, correctional facilities, Telehealth/ cyber care services, and colleges/universities;
- Holistic nurses with advanced education can become advanced practice nurses, faculty, administrators, and researchers.

Education

Nursing Programs in the US

There are 679 RN-to-Baccalaureate and 209 RN-to-Master's degree programs, with 28 new RN-to-Baccalaureate and 31 new RN-to-Master's programs under development[1(p.4)]. There are 269 Doctor of Nursing Practice (DNP) programs and 134 PhD programs.[1(p.8)]

American Holistic Nurses Certification Corporation (AHNCC)-Endorsed Academic Programs in Holistic Nursing

13 Undergraduate; 8 Graduate; 1 post Graduate certification

Nursing Accrediting Bodies

Commission on Collegiate Nursing Education (CCNE)
Accreditation Commission for Education in Nursing (ACEN)

Certifying Organizations for Holistic Nursing

American Holistic Nurses Certification Corporation (AHNCC)

Regulation and Certification

- National Licensure Exam (NCLEX) for all registered nurses through the National Council of State Boards of Nursing
- 25 (47%) State Boards of Nursing have a formal policy, position, or inclusion of holistic/complementary/integrative therapies in the scope of practice of nurses
- National Board Certification in Holistic Nursing at the basic (HN-BC – diploma and associate degree), (HNB-BC – baccalaureate degree) or advanced (AHN-BC- master's and doctoral) levels through the American Holistic Nurses Certification Corporation (AHNCC)
- American Holistic Nurses Association (AHNA) and American Nurses Association (ANA) co-published *Holistic Nursing: Scope and Standards of Practice, 2nd ed. (2013).*[2]

Research

The following are examples of the types of research being done by holistic nurse researchers.

Quantitative

Outcome measures of various holistic therapies, e.g., therapeutic touch, prayer, presence, relaxation, aromatherapy; instrument development to measure holistic phenomena; caring behaviors and dimensions; spirituality; self-transcendence; cultural competence; client responses to holistic interventions in health/illness/wellness; stress; resilience; compassion fatigue.

Qualitative

Explorations of clients' lived experiences with various health/illness/life phenomena; theory development in healing, caring, well-being, intentionality, social and cultural constructions, empowerment, health decision making; healing relationships and environments; health and wellness promotion; etc.

Challenges and Opportunities 2016–2020

Education

- integration of holistic philosophy, content, and practices into nursing curricula nationally and staff development programs
- recognition, support, and legitimization of holistic nursing practice within accreditation, regulation, licensure, and credentialing processes

Research

- identification and description of outcomes of holistic therapies/interventions, relationships, and environments
- focus on whole systems research and wellness, health promotion, and illness prevention
- funding nurses for IHM and wholeness research
- dissemination of nursing research findings to broader audiences including other health disciplines and public media

Practice

- influence and change the healthcare system to a more holistic, person centered orientation
- development of caring cultures within healthcare delivery models and systems
- collaboration with diverse healthcare disciplines to advance holistic health care
- improvement of the nursing shortage through incorporation of self-care and stress-management practices for nurses and improvement of healthcare environments

Policy

- coverage and reimbursement for holistic nursing practices and services

- education of the public about the array of healthcare alternatives and providers
- increase focus on wellness, health promotion, access, and affordability of health care to all populations
- care of the environment and the planet

Resources

Organizations

American Holistic Nurses Association (AHNA), founded 1982, the definitive voice of holistic nursing, provides vision, direction, and leadership in developing and advancing holistic philosophy, principles, standards, and guidelines for practice, education, and research. AHNA's mission is "to advance holistic nursing through community building, advocacy, research, and education". AHNA is committed to promoting wholeness and wellness in individuals, families, communities, nurses themselves, the nursing profession, and the environment. Through its various activities, AHNA integrates the art and science of nursing in the profession; unites nurses in healing; focuses on health, preventive education, and the integration of allopathic, complementary and integrative health care and healing modalities; honors individual excellence in the advancement of holistic nursing; and influences policy for positive change in the healthcare system. Phone: 800-278-2462, Email: info@ahna.org; www.ahna.org.

American Nurses Association (ANA) represents the interests of the nation's 3.2 million registered nurses (RNs) through its 55 constituent member associations, and its 37 specialty nursing and workforce advocacy organizations that currently connect to ANA as affiliates. The ANA advances the nursing profession by fostering high standards of nursing practice, promoting the rights of nurses in the workplace, projecting a positive and realistic view of nursing, and by lobbying Congress and regulatory agencies on healthcare issues affecting nurses and the public. http://www.nursingworld.org.

Citations

1. American Association of Colleges of Nursing (AACN) 2015 Annual report http://www.aacn.nche.edu/publications/annual-reports#.WEs6f-C2-p4.email
2. American Holistic Nurses Association/American Nurses Association. Holistic Nursing: Scope and Standards of Practice, 2nd ed. Silver Spring, MD: Nursesbooks.org. 2013

Homeopathy

Third Edition (2017) Editors: Rick Cotroneo, MA, CCH; Alastair Gray, BAHons MSC, DSH, PCH, PCHom; Heidi Schor, CCH, LMT, CC

Second Edition (2013) Editor: Heidi Schor, CCH, LMT, CC

First Edition (2009) Author: Todd Rowe, MD, MD(H), CCH, DHt

Second (2013) and Third Edition (2017) Partner Organization: Accreditation Commission for Homeopathic Education in North America (ACHENA)
First Edition (2009) Partner Organization: American Medical College of Homeopathy (AMCH)

About the Authors/Editors: Rowe is the Homeopathic Program Director at the Phoenix Institute of Herbal Medicine & Acupuncture and the Executive Director for the Foundation for PIHMA Research and Education. He is past President of the Arizona Board of Homeopathic and Integrated Medicine Examiners, has served on the board of directors for the Council for Homeopathic Certification and is a past President of the National Center for Homeopathy. Schor is a past President of the Accreditation Commission for Homeopathic Education in North America, a former board member of the Council for Homeopathic Certification, a past President of the Washington State Homeopathy Association, former faculty for Bastyr University, Seattle School of Homeopathy and the Homeopathic Academy of Southern California, and served as Founder and Director of the Homeopathy Community Clinic of Kirkland, a low income student-learning clinic. Gray is President of the Accreditation Commission for Homeopathic Education in North America, Academic Director of the Academy of Homeopathy Education NYC, and Director of the College of Natural Health and Homeopathy. Cotroneo is a past president and current commissioner of the Accreditation Commission for Homeopathic Education in North America.

Philosophy, Mission, Goals

Homeopathic medicine is holistic, informed by science, safe to use, and inexpensive. Homeopathy can be effective in both acute and chronic disease and can be useful in many emergency situations as well.[1] It is a system of medicine that is unique and distinct from other systems, disciplines, or modalities such as biomedicine, naturopathic medicine, herbal medicine, acupuncture, nutritional medicine, and mind-body medicine. Based on the principle of applying similars, not opposites, Homeopathy uses minute doses of natural substances to activate the body's self-regulatory healing mechanisms. It was founded by Dr. Samuel Hahnemann over 200 years ago, although the principles on which it is based have been utilized in healing for thousands of years. Since its inception, homeopathy has been used by people from all walks of life, all ages, and in countries all over the world.

Characteristics and Data

Homeopathic medicine is the second most common form of alternative medicine in the world today[2] and the most common in some higher income countries.[3] Homeopathic medicine was first introduced into the United States in 1825 where it flourished until around 1900 when the field began to meet opposition from conventional medicine.

At that time, there were 22 homeopathic medical colleges and 20 percent of physicians used homeopathic medicine. The number of current homeopathic practitioners in this country is estimated at 8500.[4] There are four subgroups practicing homeopathy in some form in the United States. Standards, regulation, licensure, and educational accreditation vary widely between these groups.

- *Lay homeopaths:* No training standards, testing or recognition.
- *Professional homeopaths:* Certified Classical Homeopath (CCH) credential offered by the Council on Homeopathic Certification; Not licensed as healthcare professionals; minimum of 1,000 hours of training, including 500 hours of didactic education and 500 hours of clinical training; no external recognition, accreditation, or licensure; national member organization is North American Society of Homeopaths (NASH). Health freedom legislation covers these practitioners in at least seven states.
- *Registered homeopathic medical assistant:* 300 hours of training required by state registration boards in Arizona and Nevada; MD or DO supervision required.
- *Licensed healthcare professionals:* Training ranges from 200–300 hours to 1000 hours, depending on the field and the level of specialization.

Scope of Practice

The scope of practice varies considerably from state to state and is defined by various licensing agencies as well as, in some states, health freedom legislation. Homeopathic licensing boards exist in Arizona, Connecticut, and Nevada for MDs/DOs; in 15 states a section of naturopathic medical board exams is on homeopathy.

Integration Activities

The *American Journal of Public Health* recently published a survey article from Harvard showing that homeopathic medicine, while still only used by a small fraction of the US population, has jumped 15% in use. In addition, most users put homeopathy among the top three complementary and integrative strategies they use in their health care. The article notes that prior studies of homeopathy "suggest potential public health benefits such as reductions in unnecessary antibiotic usage, reductions in costs to treat certain respiratory diseases, improvements in peri-menopausal depression, improved health outcomes in chronically ill individuals, and control of a Leptospirosis epidemic in Cuba."[5]

A growing number of healthcare settings offer treatment options that may include homeopathy. Homeopathic schools offer services in many offsite clinics. Examples of clinical areas of practice include fertility, acute care, early childhood, otitis media, low-income populations, substance abuse, homeless/free clinics, urgent care centers, and nursing homes. Homeopathic practitioners are sought out to provide expertise in the field of complementary medicine, including policy development, medical training, medical research, and clinical applications of natural therapies. Homeopaths Without Borders offers homeopathy volunteer services and homeopathy training to individuals living in regions facing crisis conditions. Combined degree programs are offered in homeopathic medicine and acupuncture.

Education

There are approximately fifteen homeopathic schools in the US, nine of which are either recognized or seeking recognition from the Accreditation Commission for Homeopathic Education in North America (ACHENA), the accrediting body for homeopathic schools. ACHENA is actively pursuing recognition by the US Department of Education. Naturopathic medicine is the only healthcare profession which requires homeopathic training based on Department of

Education recognized accrediting standards for all of its practitioners. Some conventional medical schools and residencies offer electives in homeopathic medicine.

Regulation and Certification

Laws regulating the practice of homeopathy vary from state to state. Usually it can be practiced legally by those whose license entitles them to practice a healthcare profession or medicine (see Scope of Practice above). Health freedom laws in some states allow the practice of homeopathy by non-licensed professionals. The Food and Drug Administration regulates the manufacture and sale of homeopathic medicines in the US. The Homeopathic Pharmacopoeia of the United States was written into federal law in 1938 under the Federal Food, Drug, and Cosmetic Act, making the manufacture and sale of homeopathic medicines legal in this country.[6] Since homeopathic remedies are sold over the counter, people in all states are free to use them for self-care at home.

Students who graduate from schools that are candidates for accreditation or are accredited by ACHENA are eligible to test for national certification through various certifying organizations, the largest of which is the Council for Homeopathic Certification (CHC). Naturopathic doctors use the CHC exam and diplomats are granted a DHANP. Conventional medical doctor (MD) and chiropractic professionals each have specialty certification exams.

Research

The homeopathic profession has been conducting research for over 200 years. Several hundred studies have been published in recent years in clinical homeopathic research. Leading organizations for homeopathic research are the Foundation for PIHMA Research and Education, Society for the Establishment of Research in Classical Homeopathy (SERCH), and Homeopathy Research Institute.

Research in homeopathy focuses on five areas:

- basic sciences research
- clinical sciences research
- educational research
- homeopathic research
- practice-based research

Challenges and Opportunities

Key challenges 2016–2020

- increasing public visibility of homeopathic medicine
- establishing widespread legality of practice
- need for expansion of homeopathic research
- opposition to regulation of the profession
- continued availability of Homeopathic Medicines
- continued improvement of educational standards for the profession
- requirement of accredited education for national certification

Key opportunities 2016–2020

- potential for the development of bridge programs in homeopathic medicine for other healthcare professionals
- increased demand for homeopathic services
- increased participation in Integrative Medicine field
- growth and professionalization of homeopathic schools
- active health freedom initiatives in multiple states
- establishment of national homeopathic medical school
- CHC attainment of Institute for Credentialing Excellence accreditation
- ACHENA attainment of Department (Secretary) of Education recognition
- inclusion in the Affordable Care Act as part of national healthcare workforce as integrative healthcare practitioners

Resources

Organizations

The following are the principal national homeopathic organizations:

- Academy of Veterinary Homeopathy (AVH)
  www.theavh.org
- Accreditation Commission for Homeopathic Education in North America (ACHENA)
  www.achena.org
- American Association of Homeopathic Pharmacists (AAHP)
  www.homeopathicpharmacy.org
- American Institute of Homeopathy (AIH)
  www.homeopathyusa.org
- Council for Homeopathic Certification (CHC)
  www.homeopathicdirectory.com
- Homeopathic Academy of Naturopathic Physicians (HANP)
  www.hanp.net
- Homeopathic Nurses Association (HNA)
  www.nursehomeopaths.org
- Homeopathic Pharmacopoeia of the United States
  www.hpus.com/overview.php
- Homeopaths Without Borders (HWB)
  www.homeopathswithoutborders-na.org
- National Center for Homeopathy (NCH)
  www.nationalcenterforhomeopathy.org
- North American Network of Homeopathic Educators (NANHE)
  www.homeopathyeducation.org
- North American Society of Homeopaths (NASH)
  www.homeopathy.org

National Center for Homeopathy (NCH) is the national consumer association, North American Society of Homeopaths (NASH) is the national association for homeopathic practitioners. Accreditation Commission for Homeopathic Education in North America (ACHENA) is the association for recognition of schools. The Council for Homeopathic Certification (CHC), Homeopathic Academy of Naturopathic Physicians (HANP) and American Board of Homeotherapeutics (ABHT) are the main national certification bodies for homeopathic practitioners.

Citations

1. Comparing homeopathy with conventional medicine. Homeopathy Research Institute website. https://www.hri-research.org/resources/homeopathy-the-debate/essentialevidence/conventional-medicine/
2. Homeopathy used around the world. Homeopathy Research Institute website. https://www.hri-research.org/resources/homeopathy-the-debate/essentialevidence/use-of-homeopathy-across-the-world/
3. Ong CK, Bodeker G, Grundy C, et al. World Health Organization Global Atlas of Traditional Complementary and Alternative Medicine (Map Volume). Kobe, Japan. World Health Organization, Centre for Health Development, 2005:61
4. National Homeopathic Practitioner Survey. January 2007. http://www.amcofh.org/research/community, Accessed November 17, 2009.
5. Integrative Practitioner. John Weeks March 2016. http://integrativepractitioner.com/whats-new/news-and-commentary/harvard-study-has-good-news-for-homeopathic-medicine/
6. Homeopathic Pharmacopoeia of the United States www.hpus.com/overview.php. Accessed November 17, 2009

Integrative Health & Medicine

Third Edition (2017) Authors/Editors: Mimi Guarneri, MD;
Bradly Jacobs, MD; Tabatha Parker, ND

Second Edition (2013) Editors: Molly Roberts, MD, MS;
Steve Cadwell

First Edition (2009) Authors: Hal Blatman, MD;
Kjersten Gmeiner, MD; Donna Nowak, CH, CRT

Third Edition (2017) Partner Organization:
Academy of Integrative Health & Medicine (AIHM)

First Edition (2009) and Second Edition (2013) Partner Organization:
American Holistic Medical Association (AHMA)

About the Authors/Editors: Blatman is a past Board President, Gmeiner previously served as a Trustee, and Nowak is a past Executive Director for the American Holistic Medical Association. Roberts is a past Board President, and Cadwell served as Executive Director of the American Holistic Medical Association. Guarneri is President, Jacobs is Board Chair, and Parker is the Associate Director of Special Academic Programs at the Academy of Integrative Health & Medicine.

Philosophy, Mission, and Goals

The Academy of Integrative Health & Medicine (AIHM) emerged in 2013 as a solution to our growing health care crisis. The formation of the AIHM is a bold response to a call to action for change — to transform the way we think about health and health care. The Academy is home to a broadening international community of healthcare practitioners, health seekers, and advocates connected by a shared holistic philosophy of person-centered care. The AIHM community recognizes the link between our health and the state of the planet and deeply considers the social determinants of health and the quality of our relationships to ourselves and one another.

One irony of the integrative healthcare movement is that while there is an increasing abundance of integrative practitioners, they often operate separately from one another. Interprofessional communication, collaboration, education and team-based care are at the heart of the AIHM's mission as we unite the many voices in integrative health and medicine to transform health care together.

Mission

The AIHM is dedicated to engaging a global community of health professionals and health seekers in innovative education, training, leadership, interprofessional collaboration, research, and advocacy that embraces all global healing traditions, to promote the creation of health and the delivery of evidence-informed comprehensive, affordable, sustainable person-centered care.

Vision

We, too, have a dream.
Where healthcare is about health and available to all
Where prevention is our foundation and mechanical fixes are embraced when we need them
Where all healthcare providers work collaboratively to heal body, mind and spirit
Where sustainability is integrated into our culture, practice and training
We are the solution
Working together to establish a new paradigm of health care for humanity and the planet

Historically and affectionately called 'the beach house meeting", twelve individuals met to transform the holistic community into a more collaborative interprofessional, international organization. They represented the American Board of Integrative Holistic Medicine (ABIHM), the Academic Collaborative for Integrative Health (ACIH) the Academic Consortium for Integrative Medicine and Health (ACIMH), the American Holistic Medical Association (AHMA), the American Holistic Nurses Association (AHNA), Healing Beyond

Borders (HBB), the Integrative Healthcare Policy Consortium (IHPC) and leaders in the naturopathic physician community.

The ABIHM and the AHMA worked together for decades to advance holistic medicine in a variety of ways. For fifteen years, the ABIHM was the leading organization to deliver a certification program in integrative holistic medicine to MD and Doctors of Osteopathic (DO) physicians. This work helped to inspire the development of the national board certification in Integrative Medicine now being offered by the American Board of Physician Specialties. The AHMA, which was founded in 1978, helped to transform the conventional/allopathic medical system to a more holistic model.

The merging of the two organizations, ABIHM and AHMA, was spearheaded by Dr. Mimi Guarneri and Dr. Molly Roberts. The ground work for such a merger was already embedded in the cultures of both organizations under the leadership of Scott Shannon, MD and Patrick Hanaway, MD. Dr. Guarneri ignited the call to action by inviting leadership from all over the country to meet and design a truly interprofessional board dedicated to Integrative Health and Medicine. These beach house visionaries included Mimi Guarneri, MD, Scott Shannon, MD, Nan Sudak, MD, Daniel Friedland, MD, Molly Roberts, MD, Steve Cadwell, John Weeks, Len Wisneski, MD, Pamela Snider, ND, Rauni Prittinen King, RN, Gene Kallenberg, MD, and Moira Fitzpatrick ND, PhD.

The AIHM community believes that integrative health care is inherently interprofessional. Medicine is but a section of the total picture of health care, highlighting the diagnosis, treatment, and prevention of *disease*. The placement of health first is intentional— in front of medicine — and emphasizes the focus on health *first*. Integrative Health & Medicine is the future of medicine, health, and wellness. It also honors all the health professionals equally (not just health professionals that diagnose or treat disease), with the patient at the center. Think of it as Integrative Medicine 2.0. It is inclusive and health-centered.

A Game Changer

The AIHM Interprofessional Fellowship offers a meaningful solution to the current silo model of integrative health care.[1] AIHM's 1,000-hour hybrid online program with residential retreats and clinical immersion experiences is the first truly interprofessional fellowship in integrative health and medicine that is open to a wide variety of health professionals. AIHM Fellows learn with and from one another and represent multiple professional disciplines, including medical, osteopathic, naturopathic and chiropractic physicians, advanced practice nurses, registered nurses and direct entry midwives with a master's degree or doctorate in a health-related field, physician assistants, licensed acupuncturists, registered dietitians, dentists, pharmacists, psychologists, and licensed clinical social workers. Fellows will become change agents for their communities and expert communicators with professionals in other disciplines, thus paving the road to elevate interprofessional education and collaboration. As the AIHM community and its influence grows, we anticipate the interprofessional model to expand dramatically, as true integration requires that we learn how to actually work together.

Characteristics and Data

Integrative medicine in the US was conceived and practiced by medical doctors as early as the 1950s, then under the name of holistic medicine, and built on various complementary medical traditions in the US dating back to the mid-19th century. Between 2000 and 2014, the American Board of Integrative and Holistic Medicine (ABIHM) provided the only peer-reviewed psychometrically validated board certification exam in Integrative Medicine which certified nearly 3,000 MDs and DOs in Integrative Medicine. In May of 2014, the American Board of Integrative Medicine (ABOIM) replaced the ABIHM as the defining, recognized board certification in Integrative Medicine. The ABOIM was formed by the American Board of Physician Specialties (ABPS) and as of December 2016, has certified nearly 400 MDs and DOs in Integrative Medicine.

Scope of Practice

Scope of practice for integrative medicine MDs and DOs is defined by conventional medical licensing statutes and agencies in all 50 states. Three states, Arizona, Connecticut, and Nevada, have homeopathic medical boards for medical doctors practicing homeopathy and other complementary and integrative therapies.

Education

Integrative physicians receive their graduate training in another field and then become board-certified in integrative medicine after completing a mandatory fellowship in integrative medicine. Many postgraduate fellowships are available in integrative medicine. A list of ABPS approved fellowships can be found here: http://abpsus.org/integrative-medicine-fellowships.

Regulation and Certification

With the formalization of an integrative medicine specialty through ABPS, the ABIHM formally sunset in December of 2016. To be eligible for certification in integrative medicine through the ABOIM an applicant must follow a variety of guidelines including being a graduate of a recognized college of medicine (MD/DO), holding a valid and unrestricted license(s) to practice medicine in the United States, its territories or Canada, an approved residency (by ACGME, AOA, RCPSC or CFPC), or have a current board certification granted on or before December 1, 2001, be board certified in another specialty and after 2016, have completed an ABOIM approved Fellowship in Integrative Medicine or have graduated from an accredited Council on Naturopathic Medical Education (CNME) college OR have graduated from an accredited Accreditation Commission on Acupuncture and Oriental Medicine (ACAOM) college OR have graduated from an accredited Council on Chiropractic Education (CCE) college.

Research

Studies of clinical sites practicing integrative/holistic medicine are underway to examine the impact of specific modalities and therapies. According to the Bravewell Collaborative, over the past two decades there has been documented growth in the number of clinical centers providing integrative medicine, the number of medical schools teaching integrative studies, the number of researchers studying integrative interventions, and the number of patients seeking integrative care.[2,3]

Challenges and Opportunities

Key challenges 2016-2020

- advancing the principles of integrative health & medicine into the conventional medical model
- maintaining economic stability of the organization and the field
- establishing legitimacy amid the current medical hierarchy
- establishing clear procedures for proper licensing and training

Key opportunities 2016-2020

- connecting health providers together through member gatherings and retreats
- increasing the understanding and stability of the field of integrative health and medicine through the strengthening of interprofessional relationships
- bringing integrative health and medicine to communities across the country
- creating innovative continuing educational opportunities including face-to-face, virtual and hybrid events
- increasing the number of physicians who are adequately trained and Board certified through the ABOIM

Resources

Organizations and Websites

Academy of Integrative Health & Medicine (AIHM)
6919 La Jolla Blvd., San Diego, CA 92037
Phone (858) 240-9033
www.aihm.org

Citations

1. The First Truly Interprofessional Fellowship for Integrative Clinicians. The Academy of Integrative Health & Medicine website. https://www.aihm.org/fellowship/
2. The Bravewell Collaborative. Integrative Medicine Improving Health Care for Patients and Health Care Delivery for Providers and Payors. The Bravewell Collaborative website. 2010. http://www.bravewell.org/content/pdf/IntegrativeMedicine2.pdf
3. The Bravewell Collaborative. The Efficacy and Cost Effectiveness of Integrative Medicine A Review of the Medical and Corporate Literature. The Bravewell Collaborative website. 2010. http://www.bravewell.org/content/IM_E_CE_Final.pdf

Integrative Medicine

Third Edition (2017) Editor: Margaret A. Chesney, PhD

Second Edition (2013) Editor: Benjamin Kligler, MD

First Edition (2009) Author: Victor S. Sierpina, MD

Partner Organization: Academic Consortium for Integrative Medicine and Health (formerly Consortium of Academic Health Centers for Integrative Medicine)

About the Authors/Editors: Sierpina is a Professor at UTMB Dept. Family Medicine and currently holds the W.D. and Laura Nell Nicholson Family Professorship in Integrative Medicine. He is a past Chair of the Academic Consortium for Integrative Medicine and Health (ACIMH). Kligler is the founding national Director to lead the integrative health strategy at the Coordinating Center for Integrative Health of the US Veterans Health Administration (VHA), Professor of Family and Social Medicine at Albert Einstein College of Medicine and a past Chair of ACIMH. Chesney, is Professor of Medicine at the University of California, San Francisco and is the Immediate Past Chair of ACIMH.

Characteristics and Data

Integrative medicine and health is a movement to transform medicine and health care led by a growing set of leaders from conventional academic health centers in association with community-based practitioners from diverse disciplines. The lead organization in that movement is the Academic Consortium for Integrative Medicine and Health (the Consortium, formerly the Consortium of Academic Health Centers for Integrative Medicine).

The Consortium is currently composed of over 70 North American academic health centers and health systems which have active programs in at least two of the three areas of education, clinical care, and research in integrative medicine and have support at the Dean's level or above. The Consortium defines integrative medicine and health as reaffirming "the importance of the relationship between practitioner and patient, focuses on the whole person, is informed by evidence, and makes use of all appropriate therapeutic and lifestyle approaches, healthcare professionals and disciplines to achieve optimal health and healing."

The Consortium started with eight schools in 1999 and, with the generous support of the Bravewell Philanthropic Collaborative for infrastructure, has grown rapidly expanding to 23 schools by 2003, 36 in 2006, and 54 in 2013. For a complete list of member schools and other Consortium information, see the Consortium website listed in the Resources section.

The mission of the Consortium is "to advance integrative medicine and health through academic institutions and health systems." The vision of the Consortium is "A transformed healthcare systems promoting health for all." The Consortium provides its institutional membership with a community of support for their academic missions and a collective voice for influencing change. The goals of the Consortium are:

- Support and mentor academic leaders, faculty, and students to advance integrative healthcare education, research, and clinical care.
- Disseminate information on rigorous scientific research, educational curricula in integrative health and sustainable models of clinical care.
- Inform health care policy.

With the voice of over one-third of US and Canadian medical schools, the Consortium is poised to continue to advocate within academia for a relevant role for integrative medicine, across all institutional missions. The Consortium and the Collaborative have collaborated in planning four major international research conferences from 2009 through 2016, and co-sponsored a conference on education in integrative health care in October 2012.

Leadership and Working Groups

The Consortium leadership is composed of a Board of Directors which is comprised of the Chair, Vice-Chair, Treasurer-Secretary, the immediate past Chair, and 3 At-Large Members. The Board meets monthly by phone and in person twice annually. Officers are elected by a Steering Committee (SC), which consists of one representative from each of the member institutions. The SC meets in person at the Consortium's annual membership meeting. The Consortium is governed by bylaws and is an IRS-recognized nonprofit 501(c)(3) organization.

Education Working Group

The Education Working Group (EWG) is involved in fostering projects in integrative medical and health education. Previous projects have included creating a curriculum guide containing exemplars of integrative medicine teaching, authoring an article on proposed competencies for a medical school curriculum in integrative medicine, and coordinating a grant to foster mind-body training and curriculum among the Consortium schools. With support of the Consortium leadership, the EWG formed a Fellowship Task Force to draft integrative medicine fellowship core competencies. In 2014 *Academic Medicine* published these core competencies, the process the task force followed in developing them, and associated teaching and assessment methods, recommendations for faculty development, and potential challenges and future directions. The EWG also hosts a Student Leadership Group, with student representatives from member institutions, and there will soon be a resident/fellow group as well to serve students after they move on to postgraduate training. Other current EWG projects include an effort to collect and assess existing tools for evaluating outcomes of educational interventions in integrative medicine, and another to collect currently available online teaching resources from member institutions for more effective dissemination.

A major accomplishment of the EWG has been the "Leadership and Education Program in Integrative Medicine (LEAPS into IM)" project, now entering its eighth year. This program, a collaboration with the American Medical Student Association, offers a one-week, immersion-style training program in integrative medicine leadership to 30 students from medical as well as other health professional schools from across North America.

Clinical Working Group

The Clinical Working Group (CWG) examines clinical initiatives across the member schools and shares expertise on clinical care, clinical models, and business issues, as well as fostering clinical research. In addition to the core working group, the CWG has created collaborations with several subgroups in specific clinical areas, including integrative oncology, pediatrics, pain medicine, cardiology, integrative mental health, and financial/business models. These groups meet on a regular basis via telephone, offering talks to members from leaders in each particular area. The calls are open to all Consortium members and are recorded and stored on the website for reference.

Research Working Group

The Research Working Group (RWG) has created a network of researchers from Consortium institutions to share their expertise and patient populations for recruitment in studies in integrative medicine. The RWG recently created four sub-groups to allow researchers with specific areas of interest to come together more readily for possible collaboration and collective learning. These groups, which will meet regularly by conference call, are in the areas of cancer symptomology, workplace wellness, models of integrative healthcare and personalized healthcare.

The RWG also plays a major role in planning the Consortium's Research Conference, for which members serve as contributors and scientific reviewers.

Policy Working Group

The Policy Working Group (PWG) of the Consortium informs the Board and Consortium member institutions of policy level issues that could influence the mission of the Consortium and provides information to policy makers regarding the Consortium, integrative medicine and health, including issues relevant to education, research and clinical practice. A priority of the PWG is to continue to monitor development and implementation of federal legislation and advise the Consortium of opportunities related to Integrative Medicine. The PWG also coordinates requests for testimony at Congressional briefings or other hearings pertaining to government policy.

Global Advances in Health and Medicine

The official journal of the Consortium is *Global Advances in Health and Medicine*, a peer-reviewed scholarly journal that publishes papers on integrative medicine and health, lifestyle and behavioral medicine, including basic and applied research. The journal was acquired by the Consortium in late 2016 and began publishing online in 2017.

Challenges and Opportunities 2016-2020

While the future of integrative medicine education is bright, some major challenges and opportunities lie ahead for us to prepare the next generation of integrative medicine practitioners. Among these are:

- Faculty development in both conventional and Complementary and Integrative Health educational institutions regarding the content of the field, teaching methods, and research expertise
- Interprofessional collaboration in education, research, clinical care, and healthcare policy between conventional and complementary and integrative health disciplines
- Providing learners with critical thinking skills, access to reliable resources, and role modeling of integrative practice in consistent, credible, and relevant ways
- Outcomes-based practice and optimal care pathways informed by research
- Changes in health insurance reimbursement and policy to provide broader access to integrative medicine for under-served patients

Resources

Organizations and Websites

- Academic Consortium for Integrative Medicine and Health
  www.imconsortium.org

Congresses

International Congress on Integrative Medicine and Health (May 2006, May 2009, May 2012, May 2014, May 2016). Next conference scheduled for May 2018 in Baltimore.

Bibliography

Consortium-endorsed:

Ring M, Brodsky M, Low Dog T, Sierpina V, Bailey M, Locke A, Kogan M, Rindfleisch M, Saper R. Developing and implementing core compentencies for integrative medicine fellowships. *Acad Med*. 2014; 89:421-248.

Kligler B, Maizes V, Schachter S. Core competencies in integrative medicine for medical school curricula: a proposal. *Acad Med*. 2004; 79:521-531.

Featured articles:

Academic Medicine issue with ten articles on integrative medicine educational programs in medical and nursing schools, with major involvement by multiple Consortium programs. *Acad Med*. 2007; 82.

Yoga Therapy

First Edition (2009) Author, Second Edition (2013) and Third Edition (2017) Editor:
John Kepner, MA, MBA, C-IAYT

Partner Organization: International Association of Yoga Therapists

About the author: Kepner is the executive director of the International Association of Yoga Therapists.

Philosophy, Mission, and Goals

IAYT supports research and education in yoga and serves as a professional organization for yoga teachers and yoga therapists worldwide. Our mission is to establish yoga as a recognized and respected therapy.

IAYT published a definition of the field in 2012 as part of our standards effort. In brief:

> Yoga therapy is the process of empowering individuals to progress towards improved health and well-being through the application of the teachings and practices of yoga…The practices of yoga traditionally include, but are not limited to, asana (postures), pranayama (breath control), meditation, mantra, chanting, mudra, ritual and a disciplined lifestyle.[1]

In general, yoga therapy is an orientation to the practice of yoga that is focused on healing, as opposed to fitness, wellness, or spiritual support. It is often taught privately or for special populations. More specifically, IAYT distinguishes yoga therapy from yoga teaching by the following elements:

- **Addresses specific conditions** by providing yoga therapy sessions to address identified and defined conditions, with consideration for the individual circumstances and nature of each client. It does not include teaching students/clients yoga for life stages such as pregnancy or menopause, where there are no difficulties or conditions.

- **Conducting an intake** appropriate to the individual client so that the session(s) can be focused on their specific therapeutic needs and concerns.
- **An appropriate yoga therapy intervention** is developed and delivered based on the information gathered in the intake and the assessment of the client.

Characteristics and Data

Yoga is one of the most popular Complementary and Integrative Health (CIH) modalities - used by approximately 9.5% of the adult population in 2012, rising from 5.1% in 2002 (See Fig 2.1)

The field of yoga therapy is self-regulated, primarily by the International Association of Yoga Therapists (IAYT). IAYT published Educational Standards for the Training of Yoga Therapists in 2012, accredited the first set of training programs to meet those standards in 2014 and began certifying individual yoga therapists in mid-2016.

The field is growing rapidly. As of September 1, 2016, IAYT has over 4,000 individual members and 170 member schools, up from approximately 640 and 5 respectively in 2003. IAYT has accredited 26 programs with approximately another 15 under review. Over 300 yoga therapists have been certified by IAYT so far, with over 100 applications submitted each month during a year-long grand-parenting window ending June 20, 2017.

Scope of Practice

The scope of practice varies by method and philosophy, but most commonly can be described as the management of structural aches and pains, chronic disease, emotional imbalance, stress, and spiritual challenges using the student's own body, breath, and mind. A formal Scope of Practice is published on the IAYT website.

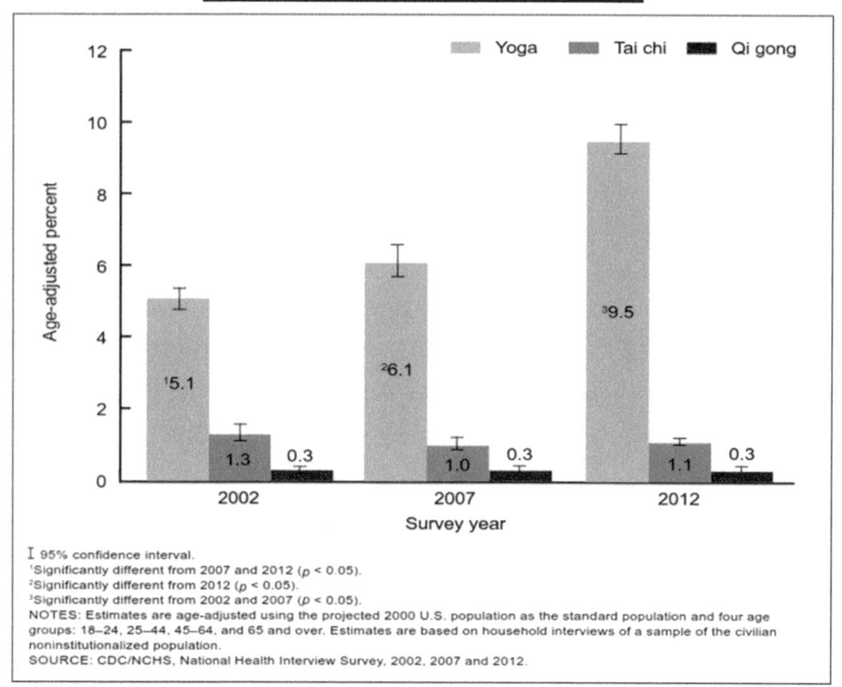

Figure 1. Use of yoga, tai chi, and qi gong among adults in the past 12 months: United States, 2002, 2007, and 2012

Integration Activities

IAYT has been a member of ACIH since 2006 and is a founding member of the Traditional World Medicine Group within ACIH. The association enjoys a close working relationship with our sister organization, the National Ayurvedic Medical Association. Yoga therapists and therapeutic yoga classes are beginning to appear in hospitals, medical clinics and mental health practices. There appears to be more and more integration with other practices, especially physical therapy, mental health services and chiropractic health care. Approximately one-third of the IAYT membership holds a health care license in other fields, with the top three being mental health, physical therapy and massage.

Education

The training of yoga therapists is most commonly provided by private schools and/or supported by traditional yoga lineages. One Master of Science degree from an accredited university is offered (Maryland University for Integrative Health) with at least two more Master's level programs on the horizon.

IAYT published voluntary educational standards for the training of yoga therapists in 2012. These are competency based standards that require at least 800 hours of training over two years after a basic 200 hour teacher training program. This is considerably more training than the standards of most yoga teacher training programs. The announcement of these standards also appears to have spurred additional interest by institutions of higher education.

Regulation and Certification

Yoga therapy is a self-regulated discipline. In North America, and, to some extent internationally, the profession is represented by the International Association of Yoga Therapists (IAYT). The IAYT standards are the highest in the western world.

Historically, individual schools provided certificates to graduates. IAYT began to provide independent credentialing in mid-2016. To date, the field has resisted outside regulation and licensing but instead has supported self-regulation through a voluntary standards process.

Research

The National Institutes of Health supports clinical research into yoga. Yoga is classified as a mind and body practice by the National Center for Complementary and Integrative Health.

IAYT is actively involved with promoting research in the field of yoga therapy. The association has produced an annual stand-alone academic yoga research conference, the Symposium on Yoga Research (SYR), since 2010. This conference was awarded a scientific conference grant by the NIH in 2011. See www.iayt.org for current information on IAYT professional research conferences.

Some members of the association are active participants in the bi-annual International Congress on Integrative Medicine and Health (ICIMH) and similar conferences. Each of the first five ICIMH conferences featured at least one symposia on yoga research and several posters on yoga research.

IAYT's journal, the *International Journal of Yoga Therapy* (IJYT), was accepted by Pub-Med in 2011.

Challenges and Opportunities

Key challenges 2016-2020

- Credentialing yoga therapists, including grandparenting current practitioners from many different lineages and from around the globe and continuing to refine the accreditation process
- Keeping the field unified, and preventing fragmentation into separate subfields stemming from other fields such as Ayurveda, medicine, mental health or physical therapy
- Developing sound models for the appropriate integration of this unlicensed field with conventional, complementary and integrative health care fields
- Financing the training required by the new educational standards, given the still emerging nature of the field and the limited opportunities for economically sustainable employment
- Supporting a global convergence towards high and comparable standards.

Key opportunities 2016-2020

The practice of yoga is a low-cost, effective practice for promoting wellness and supporting health at many levels. A key element for yoga is that, once properly trained, individuals can often practice without professional supervision.

The practice has grown tremendously over the past ten years, in part because the practice is so adaptable to the young and the old, the fit and the not-so-fit, the exercise enthusiast and the spiritual seeker. It is an especially suitable wellness practice for an aging population and shows promise in pain management, for which common medical treatments are often insufficient.

Given the growth and popularity of the field in the West in the past ten years, the emerging research supporting its many effects on various health domains, the growing acceptance of CIHM approaches in general, and the high cost of conventional care, there are many opportunities to develop the field as a stand-alone healing practice or in conjunction with conventional and CIHM healthcare fields.

Since yoga is not a licensed practice, the practice is not currently constrained by insurance reimbursement practices, although some insurers offer CIHM networks, including yoga therapy, with reduced rates for members.

Resources

Organizations and Websites

The International Association of Yoga Therapists (IAYT) is a professional association with approximately 4,000 individual members and 170 member schools in almost 50 countries (as of mid-2016). The website is www.iayt.org.

Citations

1. IAYT. Educational Standards for the Training of Yoga Therapists. IAYT website. 2012. http://www.iayt.org/?page=AccredStds

Appendices

1. About the Academic Collaborative for Integrative Health (ACIH)

2. ACIH Chronicle of Accomplishments 2004-2017

3. ACIH Acronyms

4. ACIH Board, Board Executive Committee members, Council of Advisors, Working Group Co-chairs and Staff

Appendix 1

About the Academic Collaborative for Integrative Health (ACIH)

The idea of an organization like ACIH (formerly ACCAHC – the Academic Consortium for Complementary and Alternative Health Care) initially originated in a series of meetings beginning in the late 1990s at the Integrative Medicine Industry Leadership Summits. Additional impetus was provided by the National Policy Dialogue to Advance Integrated Health Care (2001) held in Washington DC. In each instance, leading educators in integrative health professional practice recognized the value of creating a vehicle for ongoing interdisciplinary dialogue and action.

These seeds began to take formal shape in 2003 as a project of the Integrated Healthcare Policy Consortium (www.ihpc.info).

ACIH's key purpose was to bring educator leaders of the five licensed integrative health and medicine professions into one room so they could jointly articulate their shared issues and concerns about the evolving integrative medicine dialogue. ACIH provided the environment for these educators to create one voice as they joined with their academic counterparts in conventional medicine in the 2005 National Education Dialogue to Advance Integrated Health Care: *Creating Common Ground* (NED).

When these leaders held their first meeting, they realized there were many values in ongoing dialogue and collaboration. These educator leaders determined in mid-2006 that ACIH should seek to become an independent entity to imbed this movement as a continuing part of the landscape. ACIH was subsequently incorporated as a nonprofit corporation, later obtaining Internal Revenue Service recognition as a tax-exempt 501(c)(3) organization. Table A.1 lists the organizational members of ACIH, each of which pays dues based on the size of the organization.

ACIH's core members are those councils of colleges and schools, US Department of Education-recognized national specialty accrediting agencies, and nationally recognized certification and testing organizations from the licensed integrative health and medicine disciplines that choose to join. Current ACIH core membership includes 14 of these 15 organizations, plus 5 national Traditional World Medicines and Emerging Professions organizations. In addition, schools, colleges and universities affiliated with more than thirty of these accredited programs have chosen to join ACIH as individual institutional members.

As we developed our structure, we also decided not to close the door on professions that are committed to greater self-regulation but have not yet met the benchmarks of the licensed disciplines. Concerns were voiced that without such a membership provision ACIH would recapitulate the pattern of exclusion that has challenged the maturation processes of the currently licensed professions. ACIH developed distinct membership categories for two new groups, traditional world medicines (TWM) and emerging professions (EP).

Table A.1

ACIH Organizational Members as of April 2017

Member Type	Member Organizations
Councils of Colleges and Schools	Alliance for Massage Therapy Education www.afmte.org Association of Accredited Naturopathic Medical Colleges www.aanmc.org Association of Midwifery Educators http://www.associationofmidwiferyeducators.org/ Association of Chiropractic Colleges www.chirocolleges.org Council of Colleges of Acupuncture and Oriental Medicine www.ccaom.org
Accrediting Agencies	Accreditation Commission for Acupuncture and Oriental Medicine www.acaom.org Commission on Massage Therapy Accreditation www.comta.org Council on Chiropractic Education www.cce-usa.org Council on Naturopathic Medical Education www.cnme.org Midwifery Education Accreditation Council www.meacschools.org

Testing and Certification Organizations	National Board of Chiropractic Examiners www.nbce.org National Certification Board for Therapeutic Massage and Bodywork www.ncbtmb.org National Certification Commission for Acupuncture and Oriental Medicine www.nccaom.org North American Board of Naturopathic Examiners www.nabne.org
Traditional World Medicines and Emerging Professions*	Accreditation Commission for Homeopathic Education in North America http://www.achena.org/ Board for Certification of Nutrition Specialists https://nutritionspecialists.org/ Council for Homeopathic Certification http://www.homeopathicdirectory.com/ International Association of Yoga Therapists www.iayt.org National Ayurvedic Medical Association www.ayurveda-nama.org

*The Traditional World Medicines and Emerging Professions categories were created to allow participation of professions engaged in the educational and professional regulatory efforts described earlier

As mentioned above, ACIH reflects collaboration with a growing list of donors. Membership funds roughly sixty percent of our basic operations with the remainder of operational funding and all major projects dependent on foundations, corporate, and individual philanthropy. We also plan to create additional funding through the provision of resources that assist us in fulfilling on our vision. Donors that have assisted ACIH significantly in advancing our mission include ACIH's founding philanthropic partner Lucy Gonda, the Westreich Foundation, the Lia Fund, the NCMIC Foundation, the Leo S. Guthman Fund, Standard Process, Inc., Bastyr University, Life University, Andrew Weil, MD, and the Academic Consortium for Integrative Medicine and Health (formerly CAHCIM – Consortium of Academic Health Centers for Integrative Medicine). The Institute for Alternative Futures provided us with significant in-kind support. Finally, ACIH is deeply grateful for the visionary leaders of the Integrated Healthcare Policy Consortium out of whose multi-disciplinary, collaborative work ACIH was born and then nurtured in its first years.

ACIH's projects are led through the Board of Directors, the Executive Committee, and four working groups that focus on education, research, clinical care, and policy. The projects are diverse, and each includes an emphasis on supporting interprofessional education, research, and/or clinical practice. Among current or recent projects are:

- developing the ACIH *Competencies for Optimal Practice in Integrated Environments*;
- engaging in productive dialogues with both the National Institutes of Health (NIH) National Center for Complementary and Integrative Health (formerly NCCAM – National Center for Complementary and Alternative Medicine) over that agency's 2016-2020 strategic plan, and with the Patient Centered Outcomes Research Institute (PCORI) in its initial priorities, and being involved with educational institutions that have PCORI grants;
- engaging the emerging movement for interprofessional practice and education via diverse forums and relationship building including participation in the Center for Interprofessional Practice and Education at the University of Minnesota;
- completing a project with UCLA Center for Health Policy senior fellow Michael Goldstein on the role of licensed integrative health and medicine disciplines in meeting the nation's primary care needs; and,
- co-sponsoring the hugely successful International Congress for Educators in Complementary and Integrative Medicine at Georgetown University in October 2012 with our colleagues in the Academic Consortium for Integrative Medicine and Health.

Please see our Chronicle of Accomplishments in Appendix 2 for more detail on our various collaborations.

This level of engagement stimulated significant interest in ACIH playing a role in skill development for leadership to bring the values, practices and disciplines associated with integrative health care into the broader dialogue over the future of medicine and health.

Of particular note is ACIH's work with three significant projects of the Health and Medicine Division, formerly Institute of Medicine, (IOM), now under the National Academy of Medicine. The first ACIH engagement was to support consumer interest in distinctly licensed integrative health and medicine practitioners during planning and execution of the then Institute of Medicine's (IOM) February 25–27, 2009 National Summit on Integrative Medicine and the Health of the Public. Conversations between leaders of ACIH and representatives of the HMD led to the appointment of ACIH then chair and now Executive Director Elizabeth Goldblatt, PhD, MPA/HA to the HMD's planning committee. Goldblatt's service included helping form valuable links between the conventionally trained practitioners and educators who were her colleagues on the committee and key professionals in the licensed integrative health and medicine disciplines. The rich resources created through the Summit are available through the HMD's site at http://www.national academies.org/hmd/.

Subsequently, the HMD responded affirmatively to a suggestion from ACIH that a committee that was formed to develop a national strategy for pain research, education and care should include at least one representative from the licensed integrative health and medicine disciplines. We made the case that many consumers use chiropractors, acupuncture and Oriental medicine practitioners, massage therapists, Yoga therapists and naturopathic doctors for pain-related conditions. An ACIH nominee, Rick Marinelli, ND, LAc, a past president of the American Academic of Pain Management, was selected. We were pleased to see that the HMD report, "Relieving Pain in America: A Blueprint for Transforming Prevention, Care, Education, and Research," promoted an integrated, interprofessional strategy and directly referenced the contributions of complementary and integrative therapies and practitioners.

ACIH became a sponsoring organization of the 2012-2017 HMD Global Forum on Innovation in Health Professional Education. Dr. Goldblatt is the HMD Global Forum member, with John Weeks also active as an alternate. Among positive outcomes from the early engagement is a general understanding that optimal team care will involve "widening the circle" of professionals and stakeholders who are considered part of these teams. ACIH's Board charged the Goldblatt-Weeks team to work on elevating consideration of innovations that lead to education of healthcare professionals who can not only react to and manage disease but also partner with patients and communities to promote practices and choices that will lead to health and wellness. Meantime, ACIH is developing efficient means of translating the leading-edge thinking and content to its councils of schools and their individual school and program members.

More information on ACIH's areas of work, and accomplishments to date, are listed in Appendix 2, ACIH Accomplishments. Key Personnel are listed in Appendix 4. Up-to-date information on ACIH's directions and key personnel are on the ACIH website at https://integrativehealth.org/. We welcome your queries made through info@integrativehealth.org.

Appendix 2

ACIH Chronicle of Accomplishments 2004-2017*

*The ACIH Chronicle of Accomplishments with links to resources is online at https://integrativehealth.org/s/ACIH-Chronicle-of-Accomplishments_050817.pdf

2004

The Academic Collaborative for Integrative Health (ACIH) began March 2004 as an initiative of the Integrative Health Policy Consortium (IHPC). The organization was originally named the Academic Consortium for Complementary and Alternative Health Care (ACCAHC), the current name being adopted in 2016. ACIH formed shortly after several conventional academic health centers organized the Consortium of Academic Health Centers for Integrative Medicine (CAHCIM, recently renamed the Academic Consortium for Integrative Medicine and Health (ACIMH). ACIH was created, via philanthropic investment from Lucy Gonda, to 1) facilitate collaboration between educators from conventional and IHM institutions and 2) advance the shared vision and mission of IHM academic organizations. *The National Education Dialogue to Advance Integrated Health Care: Creating Common Ground (NED)* initiative was established (with support from Ms. Gonda) to facilitate and engage discussion between integrative academic leaders including ACIMH and ACIH.

ACIH launches and Pamela Snider ND becomes Founding Executive Director Pamela Snider, ND, IHPC Vice Chair, and former Bastyr University Associate Dean becomes ACIH Executive Director and Co-Founder.

Academic Leaders and Organizations From Diverse Professions to Build ACIH Executive director Pamela Snider, working with John Weeks, director of IHPC's NED, began interviewing and conferring with selected representatives of diverse disciplines, focusing on academic leaders from the five licensed integrative health and medicine professions and national academic organizations of chiropractic, acupuncture and Oriental Medicine, naturopathic medicine, massage therapy and direct-entry midwifery, all recognized by the US Department of Education. These five disciplines and representatives from their councils of colleges, accreditation commissions/agencies and certification/testing agencies formed the foundational board members of ACIH. From its inception, the ACIH Board also included emerging professions and traditional medicines to be inclusive and support professional formation. These interviews formed the basis of ACIH's vision, mission, values and strategic goals.

2005

Founding ACIH Strategic Meeting was held at Southern California University of Health Sciences, where Dr. Reed Phillips was president. Twenty academic leaders from 7 disciplines and 18 organizations were invited to work on themes for NED and to further explore and plan ongoing collaboration and development of ACIH. This was the inaugural meeting of a series of interprofessional face-to-face Board Retreats in which members of these institutions and professions would learn from, about and with each other in a classic interprofessional manner; and engage in organization planning and collaborative initiatives. These retreats have been organized regularly since the first meeting in Feb 2005: May 2006, October 2007, July 2008, May 2009, May 2010, June 2011, May 2012, October 2012, June 2013, June 2014, June 2015, and May 2016. These meetings fuel the growth and activity of the 19 national organizations and by 2015, an additional 41 individual accredited programs, colleges and universities are committed to continuous collaboration as dues-paying members of this shared collaboration.

ACIH adopts vision, mission, values, goals, structure and strategic priorities in February 2005; begins two-year organization plan.

Elections held: Founding Executive Committee is Reed Phillips, PhD (Founding Chair), Liza Goldblatt, PhD, MPA/HA, David O'Bryon JD, CAE, Jan Schwartz, MA, LMT, Don Warren ND, Pamela Snider ND (ED), John Weeks.

Interprofessional Definition of Integrative Medicine for ACIMH. A team of ACIH Board members collaborates with members of the ACIHM to change the definition of "integrative medicine" to an interprofessional definition that better respects the integration of not just modalities but also of "healthcare professionals and disciplines."

Delphi Survey of CAM Educators conducted with Oregon Collaborative for Complementary and Integrative Medicine (OCCIM) R-25 co-investigator group on Modified Delphi Survey of CAM Educators (2005). Results submitted to *JACM* (2006) *Response to a Proposal for an Integrative Medicine Curriculum*, a survey conducted by ACIH in response to Competencies in Integrative Medicine by Kligler et al, *Academic Medicine*. Publication of peer-reviewed article underscored importance of integrating disciplines and practitioners, not just therapies in "integrative medicine." Benjamin et al, *JACM*. November 1, 2007, 13(9): 1021-1034. Survey data is presented at NED.

National Education Dialogue (NED) ACIH Board members led multiple projects and task forces, participated extensively in planning for the NED (2004-2006), an initiative that involved over 60 integrative health and medicine educators from 12 disciplines that focused on ACIH leaders and those of the ACIMH. These included development of model interprofessional curricula, a Delphi survey, definitions and values, a survey on the extent of interdisciplinary relationships, and reports on best practices published in the NED Progress Report (2005). The NED was funded substantially by Lucy Gonda, directed by Weeks working closely with Snider and ACIH Task Force leaders Jan Schwartz, MA, LMT, Liza Goldblatt, PhD, MPA/HA, Dan Sietz, MS, and Patricia Benjamin, PhD.

2006

Research Conferences, Ongoing Dialogue with Integrative Medicine Educators established ACIH was a Participating Organization for ACIHM's May 2006 North American Research Conference on Complementary and Integrative Medicine (NARCCIM). ACIH leaders served on key planning committees, and submitted and delivered multiple programs and posters on ACIH surveys and initiatives. Co-hosted first ACIH/ACIHM face to face gathering between ACIHM Steering Committee and ACIH Executive Committee, 20 members gathered and acknowledged the importance of ongoing relationship at this reception.

ACIH secures three year ACIH Membership and Dues commitments from eighteen ACIH organizations going forward.

Shift to Independence ACIH leaders, with IHPC's blessing, chose to begin the process of transforming from being a project of IHPC to an independent 501c3 organization. The step was powered by a $100,000 commitment over four years from ACIH's founding philanthropic partner Lucy Gonda.

Presentations and Publications Since 2006, ACIH's projects, and research have been the subject of over two dozen papers and presentations in journals and scientific meetings. See the Presentations, Posters and Publications under 'Resources' on our website: https://integrativehealth.org.

2007

John Weeks Becomes Executive Director NED director and ACIH co-founder Weeks, the publisher-editor of the *Integrator Blog News & Reports*, became ACIH's second executive director.
Published Response on "Competencies in Integrative Medicine" Publication of a peer-reviewed paper that underscored the importance of integrating disciplines and practitioners, not just therapies, in

"integrative medicine". (Response to a Proposal for an Integrative Medicine Curriculum, Journal of Alternative and Complementary Medicine. November 1, 2007, 13(9): 1021-1034).

Researching Competencies for Integration In a grant-funded project from the National Certification Commission for Acupuncture and Oriental Medicine (NCCAOM), ACIH carried out and subsequently reported two surveys on competencies of licensed acupuncturists for practice in MD-dominated settings. This initiated a core engagement of ACIH around emerging needs of educators and practitioners in integrated settings.

2008

ACIH Incorporated and Elizabeth Goldblatt Becomes Chair ACIH incorporated as a separate 501c3 organization. Leading AOM educator Elizabeth A Goldblatt, PhD, MPA/HA, the former president of Oregon College of Oriental Medicine and past chair of the Council of Colleges of Acupuncture and Oriental Medicine became ACIH's second chair.

Organized Interprofessional Working Groups ACIH set policies to maintain balanced interprofessionalism in organizing the three multidisciplinary committees that are a center of ACIH activity. These are the Education Working Group, the Clinical Working Group, and the Research Working Group. Beth Rosenthal, PhD, MBA, MPH, ACIH Assistant Director, focuses her work supporting these groups.

Organized Interprofessional Special Interest Groups (SPIGs) In order to foster mutual understanding and to enhance the abilities of our distinct organizations, ACIH convened SPIGs among its members in these interest areas: Councils of Colleges, Accreditation Agencies, Certification and Testing Organizations and Traditional World Medicines/Emerging Professions.

Supporting the Emergence of Yoga Therapists The leaders of the ACIH Traditional World Medicine (TWM) member International

Association of Yoga Therapists (IAYT) began a multi-year process of utilizing the skills and experience of numerous ACIH personnel and member organizational leaders' expertise as resources in their campaign to create council of schools, accreditation and certification structures. The IAYT-ACIH relationship has exemplified the ACIH intent in including emerging integrative health professions that are engaging in self-regulatory and regulatory processes.

2009

Joint ACIH Meeting with Executives and Working Groups of ACIMH/ACIHM Since its founding, ACIH has fostered collegial relationships with ACIHM/ACIMH. An early highlight post-NED was a May 2009 joint executive committee and working group gathering of roughly 40 people from each organization followed by a reception and dinner. Various ACIMH leaders are on ACIH Council of Advisers.

Clarified Priority Projects ACIH board and working groups met in a strategic planning retreat at Northwestern Health Sciences University to identify core, shared, multi-year projects. Three key areas were identified: 1) supporting students, educators, clinicians and other stakeholders in working optimally in integrated environments; 2) supporting educators in developing more evidence-informed education (via a collaboration with schools that received NIH-NCCIH R-25 grants to enhance evidence-based programs); and 3) engaging the national dialogue to foster interprofessional education/care.

Publication of a Critically Needed Educational Tool ACIH, in collaboration with volunteer author teams, mainly from its Councils of Colleges member organizations, created the *Clinicians' and Educators' Desk Reference on the Licensed Complementary and Alternative Health Care Professions* to support all practitioners and educators in understanding the value of the whole practices of ACIH's professions. This fundamental interprofessional tool, with a second edition in 2013 and a third edition in 2017, has been widely heralded across multiple disciplines.

Impact on IOM's Summit on Integrative Medicine ACIH secured an invitation from the Institute of Medicine (IOM, now National Academy of Medicine) staff to ensure that the ACIH professions linked to 350,000 licensed practitioners would be represented on the Planning Committee for the February 25-27, 2009 Summit on Integrative Medicine and the Health of the Public. (www.iom.edu/Reports/2009/Integrative-Medicine-Health-Public.aspx) The meeting was funded through the Bravewell Collaborative. The ACIH nominee appointed, chair Elizabeth Goldblatt, PhD, MPA/HA, was the only person from non-MD/nursing professions on the planning team. The meeting was significantly more multidisciplinary and inclusive in part through Goldblatt's representation of patient interest and the full scope of practitioners involved in integrative health and medicine. The ACIH participation helped move the focus from "integrative medicine" toward the more inclusive "integrative health."

2010

Influential Participation in the IOM Pain Committee ACIH secured an invitation from the Institute of Medicine to ensure that ACIH member organizations would be represented on the IOM Committee on Pain Research, Care and Education. Pain expert Rick Marinelli, ND, MAOM, was nominated by ACIH. Marinelli played an important role in placing multiple references to integrative health content in the 2011 Relieving Pain in America: A Blueprint for Transforming, Prevention, Care, Education and Research and for being among the first of the committee members to articulate the need for a "cultural transformation in the way pain is perceived, judged and treated."

Workforce: ACIH's Primary Care Project In a partnership led by UCLA's Michael Goldstein, PhD and John Weeks, ACIH began a collaboration with leaders of its councils of colleges for chiropractic, naturopathic medicine, acupuncture and Oriental medicine and midwifery to explore creating a resource for all stakeholders on the substantial roles of these typically first-access professions in meeting the nation's primary care needs.

National PAINS Initiative ACIH was invited to represent patient interest in integrative care in the Pain Action Alliance to Implement a National Strategy (PAINS) led by the Kansas-based Center for Practical Bioethics. ACIH leaders Vince De Bono, DC and then Martha Menard, PhD, LMT represented ACIH on the PAINS Steering Committee and participated in regional meetings, focusing attention on the role of non-pharmacological approaches.

Development of the NIH NCCAM 2011-2015 Strategic Plan The ACIH Research Working Group (RWG) led by Greg Cramer, DC, PhD (L) and Carlo Calabrese, ND, MPH (R) led ACIH into extensive dialogue with the NIH NCCAM (now NCCIH) on the agency's strategic plan. ACIH focused on the "real world research" that is likely to most inform the ability of consumers and other stakeholders considering services from these disciplines. In letters, phone conferences and a face-to-face meeting, ACIH underscored the importance of looking at whole practices, at cost issues and in building the capacity of these disciplines to participate in the research endeavor.

Patient-Centered Outcomes Research Institute (PCORI) Via leadership of the Research Working Group (RWG), ACIH provided public comment on the new agency established under the Affordable Care Act. Two ACIH leaders were part of a panel on integrative health and medicine at a September 2011 Listening Session for the PCORI Board of Governors. RWG founding co-chair and member *emeritus*, Christine Goertz, DC, PhD, was appointed to serve on the PCORI Board.

Competencies for Optimal Practice in Integrated Environments ACIH engaged an 18-month process involving over 50 professionals from eight disciplines to create a document entitled Competencies for Optimal Practice in Integrated Environments.

2011

Meeting with NCCIH Leaders on National Research Strategy An interprofessional ACIH team met in February 2011 in a two-hour session at NCCIH headquarters in Bethesda, Maryland with NCCIH director Josephine Briggs, MD and members of her executive team. The focus was the research priorities of ACIH professions. ACIH has promoted whole system, real world research, cost studies, and research capacity building in the licensed integrative health disciplines. ACIH Research Working Group (RWG) Co-Chairs Carlo Calabrese, ND and Greg Cramer, DC attended, along with other RWG members.

Expansion of Philanthropic Investment Fulfillment of ACIH's mission requires continuous expansion of the organization's base of philanthropic partners. Work during the founding period of 2004-2009 was engaged via multi-year grants from ACIH's founding philanthropic partner, Lucy Gonda. In 2008, parties close to ACIH such as National University of Health Sciences, NCMIC Foundation, and Bastyr University began making critically important, multi-year Sustaining Investor grants. Life University began a few years later.

The Chicago-based Leo S. Guthman Fund made annual commitments beginning in 2008 through 2017. The Lia Fund provided five years of critical resources. In 2011, philanthropist and integrative health activist Ruth Westreich launched ACIH further. She began advising ACIH in multiple ways while also providing grant support via the Westreich Foundation.

Collaboration Across Borders III Via a philanthropic investment from Lucy Gonda, ACIH created a significant presence for the integrative health and medicine professions and movement at the most significant North American interprofessional education gathering. ACIH presented on its *Competencies*, sponsored a booth and distributed 700 copies of the ACIH *Desk Reference.*

2012

Institute of Medicine (IOM) Global Forum on Innovation in Health Professional Education In January 2012, the ACIH Board of Directors chose to become a founding co-sponsor of the Institute of Medicine Global Forum on Innovation in Health Professional Education. ACIH is the sole integrative health-focused sponsor of the forum, joining over 30 councils of colleges and professional associations from medicine, nursing, public health, pharmacy and other disciplines. Elizabeth A. Goldblatt, ACIH's Chair is the representative to the Global Forum; John Weeks is her alternate. ACIH's priority is to collaborate with the other forum members to explore innovations associated with graduating students who are expert in assisting individuals and communities toward wellness, self-care, personal empowerment, prevention and health promotion. ACIH's participation from 2012-2017 is made possible via a generous philanthropic investment from a foundation that chooses to remain anonymous.

Recognized as Supporting Member by IPEC When ACIH leaders learned in 2011 of the *Core Competencies for Interprofessional Collaborative Practice* published by the Interprofessional Education Collaborative (IPEC), ACIH leaders Mike Wiles, DC, MEd and Jan Schwartz, MA led an effort that merged the ACIH *Competencies* with this document. ACIH then secured endorsement of the new *Competencies* from the ACIH council of college members. This led to IPEC recognition of ACIH as a supporting member of its national interprofessional initiative.

International Congress for Educators in Complementary and Integrative Medicine Following an invitation from long-time adviser and promoter of interprofessional action Adi Haramati, PhD, ACIH engaged a co-sponsorship with ACIMH and Georgetown University for the first International Congress for Educators in Complementary and Integrative Medicine October 24-26, 2012. The Congress drew over 350 educators, researchers and clinicians. Multiple ACIH presentations were the subject of papers subsequently included in a report on the conference. Education and Clinical Working Group members were involved in a variety of presentations including: "Competencies for Public Health and Interprofessional Education (IPE) in the Accreditation Standards for the Licensed CAM Disciplines," "Naturopathic Residencies: Creating Interdisciplinary Education for Integrative Health Care," "Hospital Based Massage Therapy: Competencies and Standards of Practice," "Interprofessional Education (IPE) in the Complementary, Alternative, and Integrative Practice Programs: A Composite Portrait from Two Surveys, and an IPE Product Marketing Campaign," and "Integrated Clinical Services at Accredited Complementary and Alternative Health Care Academic Institutions."

Organized Participation in the National Center for Interprofessional Practice and Education (NCIPE) With the announcement the HRSA and foundation-funded NCIPE, ACIH organized its communities to sign-up and participate, posted resources and began routinely promoting webinars and other NCIPE resources to ACIH's growing list of participants. NCIPE director Barbara Brandt, PhD refers to ACIH's work in the IPE environment this way: "ACIH's work in interprofessional education is impressive, timely, a model to shine a light on."

Partner and Steering Committee of HRSA-Funded IMPriME ACIH was invited to serve as a partner organization of the HRSA-funded Integrative Medicine in Preventive Medicine (IMPriME) project of the American College of Preventive Medicine. Two ACIH executive committee members served on the IMPriME Steering Committee that developed competencies and a community of learners.

2013

Second Edition, *Clinicians' and Educators' Desk Reference* ACIH completed a significant update of its 2009 *Clinicians' and Educators' Desk Reference on the Licensed Complementary and Alternative Health Care*

Professions. The new edition includes e-book and a freely available PDF format.

Advancing the Role of ACIH Professions in *Meeting the Nation's Primary Care Needs* In a culmination of a partnership that began in 2010, ACIH produced the 80-page white paper entitled *Meeting the Nation's Primary Care Needs: Current and Prospective Roles of Doctors of Chiropractic and Naturopathic Medicine, Practitioners of Acupuncture and Oriental Medicine and Direct-Entry Midwives*. Presentations on the content have been made in multiple environments including International Congress for Educators in Complementary and Integrative Medicine (2012), UCLA (2013), and the Association of American Medical Colleges (2014). This project was led by Michael Goldstein, PhD, MPH and John Weeks, ACIH executive director.

Introducing Health and Wellbeing at the IOM Global Forum Through leadership of ACIH's Global Forum on Innovation in Health Professional Education member Elizabeth A. Goldblatt, PhD, MPA/HA, ACIH's core message in transitioning the workforce into one focused on health creation was introduced in a forum workshop on Establishing Transdisciplinary Professionalism for Health. ACIH invited integrative nursing leader Mary Jo Kreitzer, PhD, RN, FAAN as a presenter and discussion leader. ACIH has had a continuing role in promoting dialogue on the theme of self-care, resilience, health and wellbeing in health professional education theme of health.

The Project to Enhance Research Literacy (PERL) The PERL initiative is a collaboration between ACIH and the seven ACIH member institutions that received NIH R-25 education grants on evidence-informed practice and subsequently with multiple ACIH member organizations. PERL provides a huge body of resources and webinars to support ACIH educators and clinicians in engaging evidence issues more significantly in their education and practice. PERL corresponds to Competency #5, "Evidence-Based Medicine and Evidence-Informed Practice" of ACIH's *Competencies for Optimal Practice in Integrated Environments*. A key initiative in this project, funded significantly in

2014-2015 through successive one-year grants from the Josiah Macy Jr. Foundation, has been a webinar series on evidence informed practice (EIP). Project manager is Deb Hill, MS.

Working Group Series: Recorded Webinar Presentations ACIH's Research, Clinical and Education Working groups began regularly recording presentations in their meetings and then posting them at www.integrativehealth.org/conference-and-meeting-presentations/ for public use.

2014

Project for Integrative Health and the Triple Aim (PIHTA) ACIH began a significant project to support all stakeholders in seeking the best uses of the values, practices and disciplines associated with integrative health and medicine amidst the emerging values-based healthcare movement. The Project for Integrative Health and the Triple Aim corresponds to the focus of Competency #6 (Institutional Healthcare Culture and Practices) of ACIH's *Competencies for Optimal Practice in Integrated Environments*. Project Manager is Jennifer Olejownik, PhD.

Association of American Medical Colleges (AAMC) Conference Executive director John Weeks was invited by the AAMC to present on ACIH's disciplines in a plenary panel at the AAMC's 10th Annual Workforce Research Conference. Via a grant from the NCMIC Foundation, ACIH presented a poster on the PIHTA initiative, and provided copies of ACIH's white paper *Meeting the Nation's Primary Care Needs* to all 200 workforce experts in attendance.

Interprofessional Action: All Together Better Health #7 At this significant international conference on interprofessionalism, ACIH presented outcomes of an ACIH survey on the extent of inter-professional education in ACIH fields, its priorities as a functioning interprofessional collaborative, and a poster on the PIHTA initiative.

Publication of PAINS *Never Only Opioids* Policy Brief and Pain Recommendations Members of ACIH's Task Force on Integrative Pain Care, created a policy brief for the 40-organization coalition PAINS entitled *Never Only Opioids: The Case for Early Use of Non-Pharmacological Approaches and Practitioners in the Treatment of Patients with Pain.* The influential paper also included a two-page set of Recommendations for Policy Makers.

Special Project: IHM definitions, IPE in Clinical Education in ACIH Professions In 2013, the Leo S. Guthman Fund began a multi-year commitment to a special Clinical Working Group project focusing on interprofessionalism in the ACIH disciplines. The project led to a series of presentation and publications, including "A Qualitative Analysis of Various Definitions of Integrative Medicine and Health" and "The Extent of Interprofessional Education in the Clinical Training of Integrated Health and Medicine Students: A Survey of Educational Institutions," each published in *Topics for Integrative Health Care*. In the former, the authors reported the usage and context of terms occurring in IHM definitions by groups hypothesized to have the greatest current impact on the evolving field, concluding, among other things: "Leaving disciplines and health professionals out of the definition effectively leaves out the rich experience and context of the discipline, and devalues interprofessionalism." Beth Rosenthal, PhD, MPH, MBA and Anthony Lisi, DC led this project.

2015

Addressing the NIH NCCAM Advisory Council Imbalance ACIH responded to the decreasing number of members from its professions on the Advisory Council of the National Center for Complementary and Integrative Health (NCCIH). A Research Working Group team created a letter and board chair Elizabeth A. Goldblatt, PhD, MPA/HA and vice-chair David O'Bryon, JD co-signed the letter and met with the NCCIH director. The request included submission of 15 CVs of experienced researchers from ACIH fields. ACIH strongly supports the NCCIH's statement that skilled clinicians are critical to NCCIH fulfilling its public health agenda.

Opportunities for Integrative Health under the Triple Aim ACIH's Project for Integrative Health and the Triple Aim published survey results that showed leaders of integrative medicine programs associated with academic health centers are finding more inclusion from their parent health systems with shifts toward a values-based delivery system, under the Affordable Care Act. Results were reported in a commentary in *Global Advances in Health and Medicine* and in a posting of the patient experience research and advocacy organization The Beryl Institute.

Partnership with the Academy of Integrative Health and Medicine Following a full-day meeting between ACIH Executive Committee members and leaders of the Academy of Integrative Health and Medicine (AIHM), an ACIH team led by Stacy Gomes, EdD, partnered with AIHM's educators to develop educational content that will be used as part of multiple AIHM programs. The Westreich Foundation supported this work.

Partnership on the Education Track for the 2016 International Congress ACIH was invited by Academic Consortium for Integrative Medicine and Health (ACIMH), representing over 60 medical schools, to serve as its partner on the educational track for the May 2016 International Congress for Integrative Medicine and Health. Michael Wiles, DC, MEd is serving as ACIH's co-chair for the education content of the program with Elizabeth Goldblatt, PhD, MPA/HA, Mitchell Haas, DC, MA and Melanie Henriksen, ND, LAc, APRN on the Program Committee.

Post-Graduate Training (PGT) Report This report covers the status of residencies/PGT in the 5 licensed IHM disciplines (how they are managed, what organizations are involved, guidelines on how to start programs) along with ideas for increasing opportunities in IHM fields. The authors found that there are significant variations in PGTs among the IHM disciplines, and where there are opportunities, there is a shortage of number and types of opportunities. This report is meant to be a living document and brief updates will be posted on the ACIH website periodically.

Project for Inter-Institutional Education Relationships (PIERs) Members of the ACIH Clinical and Education Working groups compiled descriptions of what their institutions are doing related to inter-institutional education relationships. The reports are written to a standard template. The goal is to foster optimal inter-institutional and interprofessional relationships that bridge the gaps in education and practice between diverse integrative health and medicine (IHM) academic institutions, and between these institutions and other healthcare institutions and organizations.

Participation in the National Center for Integrative Primary Healthcare (NCIPH) The University of Arizona Center for Integrative Medicine (AzCIM), in collaboration with the Academic Consortium for Integrative Medicine and Health received a HRSA grant to establish the National Center for Integrative Primary Healthcare (NCIPH). ACIH Co-Executive Director, Elizabeth A. Goldblatt, PhD, MPA/HA is a member of the NCIPH leadership team. The ultimate goal of the NCIPH is education of interprofessional, patient/person centered teams that will utilize the principles of IH in delivering primary care. The NCIPH supports the incorporation of competency- and evidence-based Integrative Health Care (IH) curricula into educational programs in a movement toward integrative inter-professional patient care. The NCIPH is developing competencies, curricula and best practices taking into consideration the determinants of health including physical and social environment, individual health behaviors and health services as they relate to the practice of IH to affect health outcomes. Please see the center website for more information www.nciph.org.

Advocating for NIH Investment in CIM Institutions Members of ACIH's Research Working Group, led by Co-Executive Director

Martha Menard, PhD, LMT, co-authored a study of NIH investment in CIM institutions from 2010-2014. Based on data taken from the NIH RePORT database, RWG members made recommendations to the National Center for Complementary and Integrative Health for their 2016-2020 strategic plan regarding the need for continued investment in CIM institutions to build research capacity and create research infrastructure. The study and its recommendations were published in the July 2015 issue of the *Journal of Alternative and Complementary Medicine*.

Leadership Transitions In July, Elizabeth A. Goldblatt, PhD, MPA/HA and Martha Menard, PhD, LMT became co-executive directors. David O'Bryon, JD, CAE, the president of the Association of Chiropractic Colleges was elected interim chair of ACIH. Mr. O'Bryon is a founding member of ACIH and served as vice-chair since ACIH's inception. Dr. Goldblatt is also one of the founding board members of ACIH and served as the organization's chair from 2008 to 2015. She is an educator, past president of the Council of Colleges of Acupuncture and Oriental Medicine (CCAOM), and previous president of an AOM college. Dr. Menard is a clinician, researcher and educator who has held multiple leadership positions with the massage therapy community and several volunteer roles with ACIH since 2008. Since its inception in 2004, ACIH has developed into an independent, interprofessional, collaborative organization that now includes 58 member organizations and institutions from the integrative health and medicine professions and is involved in multiple collaborative projects throughout the country.

Further Philanthropic Investment The NCMIC Foundation pledged an annual $10,000 donation for 2016-2018. With this $30,000 award, the foundation's investment in ACIH since 2008 now totals over $100,000. Louis Sportelli, DC, guides the Foundation. Dr. Sportelli has been President of NCMIC Group, Inc. since 1995. Dr. Sportelli served as President of the World Federation of Chiropractic (WFC) from 1998 to 2000 and served as chair of the board of the American Chiropractic Association (ACA) from 1989 to 1990. The Foundation sponsored

ACIH leaders and the potential primary care contributions of chiropractors, naturopathic doctors, acupuncture and Oriental medicine professionals, and direct entry midwives at the Association of American Medical College's 2010 workforce meeting. The Foundation also provided support for the printing and distribution of ACIH's influential *Never Only Opioids* policy paper.

New Executive Committee and Officers Elected During its October meeting, the ACIH Board of Directors named new members to its Executive Committee and elected officers. New EC members include Stacy Gomes, EdD, MAEd; Dale Healey, DC, PhD; John Scaringe, DC, EdD; and JoAnn Yanez, ND, MPH. Officers elected are David O'Bryon, Chair; JoAnn Yanez, Vice-Chair; Kory Ward-Cook, PhD, MT (ASCP), CAE, Treasurer; and John Scaringe, Secretary. Stacy Gomes and Dale Healey will serve as members at large with Pamela Snider, ND, founding board member and former executive director. The new Executive Committee and officers will begin their terms in January 2016. Outgoing EC members Jan Schwartz, MA, BCTMB, Secretary, and Marcia Prenguber, ND, are founding members of ACIH and will remain on the board. The ACIH community is delighted that they will remain active members of the ACIH Board.

2016

A New Name – The Academic Collaborative for Integrative Health The Board of Directors at its January 2016 meeting voted to change its name from the Academic Consortium for Complementary & Alternative Health Care (ACCAHC) to the Academic Collaborative for Integrative Health (ACIH or "The Collaborative"). The Board and staff, with input from the ACIH community, had been discussing a name change for the last year. The new name evokes the essence of the disciplines ACIH represents, which includes a focus on health and well-being as well as the treatment of a wide variety of diseases and conditions that are approached from a collaborative, team-based patient/person-centered care. In addition, the organization is a functioning collaborative, as the core membership includes 18 national organizations representing councils of colleges, accreditation organizations recognized by the US Department of Education, and certification/testing agencies, as well as emerging disciplines and traditional world medicines and individual college/program members. The new name – the Academic Collaborative for Integrative Health – reflects the organization's purpose and vision.

ACIH Nominees Patricia Herman and Cynthia Price Chosen for NIH Advisory Council ACIH through its Research Working Group is delighted to announce two new members on the Advisory Council to the NIH National Center for Complementary and Integrative Health. ACIH submitted CVs of 15 top-quality candidates. Congratulations to RWG members Patricia Herman, ND, PhD and Cynthia Price, PhD, LMP!

National Center for Integrative Primary Healthcare (NCIPH) Publishes Competencies The Health Resources and Services Administration (HRSA) funded the National Center for Integrative Primary Health (NCIPH: www.nciph.org) has published competencies to support their mission. ACIH is involved through Elizabeth A. Goldblatt, PhD, MPA/HA's participation as a member of the $1.8-million project's Interprofessional Leadership Team. She has brought in ACIH teams of chiropractors, naturopathic doctors, acupuncture and Oriental medicine practitioners to inform the process. The project is sponsored by the University of Arizona Center for Integrative Medicine and the Academic Consortium of Integrative Medicine and Health. The 45-hour pilot course for all primary care providers in Integrative Health started in January 2016 and was made available to the larger community after June 2016.

Elevated Role for Health, Self-care, Wellbeing and Resilience - and ACIH! - at the Global Forum for Innovations in Health Professional Education The April 20-22, 2016 sessions at the National Academy of Sciences of the Global Forum on Innovation in Health Professional Education showed significantly more attention to the themes that the ACIH board prioritized for its now 4.5 years of work. ACIH's membership in the Forum, sponsored by the Health and Medicine Division (formerly the Institute of Medicine), is represented by the team of executive director Elizabeth A. Goldblatt, PhD, MPA/HA and former executive director John Weeks.

The April gathering included significant informal and formal engagement with health, wellbeing, and resilience - some in the context of the significant stress, burnout and suicide concerns in health professional education, and practice. JoAnn Yanez, ND, MPH, ACIH vice chair and executive director of the Association of Accredited Naturopathic Medical Colleges was present as was Beth Pimentel, ND, representing president's level member Maryland University of Integrative Health.

Is Self-care and Wellbeing in ACIH Member Accreditation Guidelines? As part of the preparation for the meeting, the ACIH team engaged a survey of accrediting agency document to see whether ACIH fields, and some others, included any reference to the importance of self-care. Both the Commission on Massage Therapy Accreditation and Council on Naturopathic Medical Education have some requirements on these most important areas.

ACIH at the 2016 International Congress for Integrative Medicine and Health ACIH was a partner organization for the May 17-20, 2016 International Congress for Integrative Medicine and Health in Las Vegas. The Congress was sponsored by our colleagues at the Academic Consortium for Integrative Medicine and Health (ACIMH). The ACIH Board met in Las Vegas on Monday, May 16th and focused on strategic planning as well as heard reports from the Samueli Institute, the ACIMH and from representatives from the University of Arizona National Center for Integrative Primary Care (www.ncipc.org). Several from the ACIH community presented at the Congress including Gerry Clum, DC, on the "The History of Discrimination in Integrative Health Care and What to Do About It; " ACIH Research Working Group RWG member James Whedon, DC on "Unwarranted Variation in Coverage for Integrative Healthcare

Services". The ACIH Executive Director was part of a pre-Congress workshop on the National Center for Integrative Primary Healthcare - Enhancing Interprofessional Integrative Health Education. RWG member Patricia Herman, ND, PhD, co-led a discussion on "Policy and the Economics of Integrative Medicine" and Meg Jordan, PhD with ACIH Associate Member California Institute of Integral Students was part of the panel "Mindfulness and Relational Neuroscience: The Real Foundation for Health Coaching." Belinda (Beau) Anderson, RWG member, Liza Goldblatt, Stacy Gomes, ACIH Executive Committee member and Ben Kligler, ACIH Council of Advisors presented on the "Application of ACIH Competencies for Optimal Practice in Integrated Environments: Lessons." ACIH Hospital Based Massage Therapy (HBMT) "Hospital Based Massage Therapy: Competencies and Standards of Practice" was presented by Dale Healey, MK Brennan, Carolyn Tague and ACIH Assistant Director Beth Rosenthal. Liza Goldblatt chaired plenary session 05 and introduced the plenary speaker Dr. Barbara Brandt, a national leader in IPE/CP. John Weeks, former ACIH Executive Director presented on "The Movement toward Values-Based Medicine and the Triple Aim: Is it Increasing Interest in Integrative Health Services.?" Patricia Herman also presented on "Determining Appropriateness of Manipulation and Mobilization for Chronic Low Back and Cervical Pain in the Patient Centered Care Era" and James Whedon on "Who Needs Quality Measurement for Integrative Healthcare? Lastly, ACIH was invited to be part of the Final Plenary Session "Future Directions: Looking Forward Together." This was the first time ACIMH included four areas – education, clinical, research and policy at one Congress.

Credentialing AOM Professionals In November 2016, the National Certification Commission for Acupuncture and Oriental Medicine (NCCAOM) and ACIH completed a one-year project on a resource to assist medical delivery organization to credential acupuncture and Oriental medicine (AOM) professionals. **"Credentialing AOM Professionals for Practice in Healthcare Organizations"** was completed in fall of 2016 and is now available on the ACIH web site. Authors are John Weeks, Stacy Gomes, EdD, ACIH Executive

Committee member and Elizabeth A. Goldblatt, ACIH Executive Director with NCCAOM team of Iman Majd, MD, EAMP (LAc) and CEO Kory Ward- Cook, PhD, CAE. This document includes inputs from 27 AOM professionals and delivery organization administrators who have been responsible for credentialing members of this profession.

2017

NCIPH Course Available The Academic Collaborative for Integrative Health's Executive Director, Liza Goldblatt, is on the leadership team of the **National Center for Integrative Primary Healthcare** (www.nciph.org). In January 2017, NCIPH announced that the 45-hour interprofessional integrative health online curriculum, called Foundations in Integrative Health (FIH), for primary care educational programs is currently open to **educational training programs.** In late February or early March 2017, the course will be open to **individuals.** The individual enrollment option will allow health professionals, educators and students not affiliated with an educational training program to enroll. Continuing education credit is available for the individual enrollment option (including CME, CNE, and CE for Pharmacy). Both versions of the course (either for an educational program, or for an individual), will be available online free of charge through **August 31, 2017.**

ACIH announces the National Academies Health and Medicine Division Global Forum on Innovations for Health Professional Education will hold a 1.5-day public workshop in spring 2018 ACIH's role at the Forum is two-fold 1) to educate the over 50 national health professional organizations that are members of the Global Forum about the disciplines that ACIH represents and about integrative medicine and health, and 2) to encourage health professional educators to put more emphasis on disease prevention and on creating health and well-being. We are delighted that in the *spring of 2018, the Global Forum in collaboration with the National Academy of Medicine will hold a 1.5-day public workshop on* ***"A Systems Approach to Alleviating Work-induced Stress and Improving Health, Well-being, and Resilience of Health Professionals Within and Beyond Education."***

The 1.5-day public workshop will explore systems-level causes and downstream effects of job-related stress affecting all health professions working in learning environments both clinical and classroom settings. On May 18, 2017, ACIH held a webinar on the Global Forum and the upcoming workshop with Patricia Cuff, MS, MPH, Mary Jo Kreitzer, PhD, RN, Sandeep Kishore, MD, PhD and Elizabeth Goldblatt, PhD, MPA/HA.

Appendix 3

ACIH Acronyms

Councils of Colleges

AANMC - Association of Accredited Naturopathic Medical Colleges
ACC- Association of Chiropractic Colleges
AFMTE – Alliance for Massage Therapy Education
AME – Association of Midwifery Educators
CCAOM - Council of Colleges of Acupuncture and Oriental Medicine

Accrediting Agencies

ACAOM – Accreditation Commission for Acupuncture and Oriental Medicine
CCE - Council on Chiropractic Education
CNME - Council on Naturopathic Medical Education
COMTA - Commission on Massage Therapy Accreditation
MEAC - Midwifery Education Accreditation Council

Testing and Certification Organizations

NABNE - North American Board of Naturopathic Examiners
NARM – North American Registry of Midwives
NBCE – National Board of Chiropractic Examiners
NCBTMB - National Certification Board for Therapeutic Massage and Bodywork
NCCAOM - National Certification Commission for Acupuncture and Oriental Medicine

Emerging and Traditional Professions

ACHENA - Accreditation Commission for Homeopathic Education in North America
CHC - Council for Homeopathic Certification
IAYT - International Association of Yoga Therapists
NAMA - National Ayurvedic Medical Association

YA - Yoga Alliance

Disciplines

AOM – Acupuncture and Oriental Medicine [LAc or Lic.Ac. are licensed, DAOM is doctorate in AOM, MAOM is Masters in AOM]
DC – Chiropractic
DEM – Direct-entry Midwifery
EP – Emerging Profession
IHM – Integrative Health and Medicine
IM – Integrative Medicine
MT - Massage Therapy
ND – Naturopathic Medicine
TWM – Traditional World Medicine

Others Acronyms

Within ACIH
AA SPIG - Accreditation Agency SPecial Interest Group
CC SPIG – Councils of Colleges SPecial Interest Group
CENTER – Center for Optimal Integration: *Creating Health*
CEDR – Clinicians and Educators Desk Reference (book)
CT SPIG – Certification/Testing SPecial Interest Group
EC – Executive Committee
HSCP – Hotspots/Cooling Point
PERL – Project to Enhance Research Literacy
PIHTA – Project for Integrative Health and the Triple Aim
SPIG – SPecial Interest Group

Outside ACIH
AAMC – American Association of Medical Colleges
ABIHM - American Board of Integrative Holistic Medicine
ABOIM - American Board of Integrative Medicine
ACPM – American College of Preventive Medicine
ACIMH – Academic Consortium for Integrative Medicine and Health (formerly CAHCIM – Consortium of Academic Health Centers for Integrative Medicine)

AHMA – American Holistic Medical Association
AHNA - American Holistic Nurses Association
AIHM – Academy of Integrative Health and Medicine
AMTA - American Massage Therapy Association
ATBH – All Together Better Health
CAAM- California Association of Ayurvedic Medicine
CAB III/IV – Collaboration Across Borders
CAR- Council for Ayurvedic Research
FPD – First Professional Doctorate
HMD Health and Medicine Division, formerly Institute of Medicine, (IOM), now under the National Academy of Medicine)
HRSA – Health Resources Services Administration
ICIMH – International Congress for Integrative Medicine and Health (formerly ICC-CIM International Congress for Clinicians in Complementary and Integrative Medicine and ICE-CIM International Congress of Educators in Complementary and Integrative Medicine)
IHPC – Integrative Healthcare Policy Consortium
IMPriME – Integrative Medicine in Preventive Medicine
IM4US – Integrative Medicine for the Underserved
IOM Institute of Medicine (as of March 16, 2016 is now HMD: now under the National Academy of Medicine)
IPE – Interprofessional Education
IPE/C - Interprofessional Education/Care
IPEC – Interprofessional Education Collaborative
IRCIMH – International Research Congress on Integrative Medicine and Health
MTF – Massage Therapy Foundation
NCCIH – National Center for Complementary and Integrative Health (formerly NCCAM – [National Institutes of Health] National Center for Complementary and Alternative Medicine)
NCIPE – National Center for Interprofessional Practice and Education
NED – National Education Dialogue to Advance Integrated Health Care
PCORI – Patient Centered Outcomes Research Institute

Appendix 4

ACIH Board, Board Executive Committee members, Council of Advisors, Working Group Co-chairs and Staff

ACIH Board of Directors

Karen Bobak, DC, EdD
Stanley Dawson, DC, LMBT
Courtney Everson, PhD
Iman Majd MD, MS, EAMP/LAc
Mark McKenzie, LAc, MsOM
William C. Meeker, DC, MPH
Steffany Moonaz, PhD
Paul Morin, DC
Beth Pimentel, ND
Marcia Prenguber, ND
Nichole Reding, MA, CPM, LDM
Jan Schwartz, MA, BCTMB
Pamela Snider, ND
Elizabeth A. Goldblatt, PhD, MPA/HA, ACIH Executive Director, ACIH Board ex-officio

ACIH Board Executive Committee

David O'Bryon, JD, CAE, Chair
JoAnn Yanez, ND, MPH, Vice Chair
John Scaringe, DC, EdD, Treasurer
Dale Healey, DC, PhD, Secretary
Stacy Gomes, EdD, MA Ed, Member-At-Large

ACIH Council of Advisors

Clement Bezold, PhD
Margaret Chesney, PhD
John H.V.Gilbert, CM, PhD., LLD (Dalhousie), FCAHS
Aviad (Adi) Haramati, PhD
Ping Ho, MA, MPH
Bradly Jacobs, MD, MPH
Wayne Jonas, MD
David L. Katz, MD, MPH, FACPM, FACP, FACLM
Benjamin Kligler, MD, MPH
Lori Knutson, RN, BSN, HN-BC
Mary Jo Kreitzer, PhD, RN, FAAN
Anne Nedrow, MD, MBA
Adam Perlman, MD, MPH
Sarita Verma, LLB. MD, CCFP
Ruth Westreich
Leonard Wisneski, MD

ACIH Working Group Co-Chairs

Clinical Working Group:
Marcia Prenguber, ND
Barry C. Wiese, DC, MSHA, DIBCN

Education Working Group:
Dale Healey, DC, PhD
Beth Howlett, DAOM, LAc

Research Working Group
Belinda Anderson, PhD, LAc
James M. Whedon, DC, MS

ACIH Staff

Elizabeth Goldblatt, PhD, MPA/HA, Executive Director
Beth Rosenthal, PhD, MBA, MPH, Assistant Director
Renée Motheral Clugston, Director of Operations
Deb Hill, MS, Project Manager PERL
Jennifer Olejownik, PhD, MS, Project Manager PIHTA
Carla Wilson, PhD, DAOM, LAc, Development/Program Consultant

This list is current as of April 2017; please check www.integrativehealth.org for updates

For information about ordering additional copies of this book, please visit:

www.integrativehealth.org